Improving Performance in Service Organizations

Improving Performance in Service Organizations

How to Implement a Lean Transformation

Joyce A. Miller
KeyStone Research Corporation

Tania Bogatova
KeyStone Research Corporation

Bruce Carnohan
Carnohan Process Solutions

LYCEUM
BOOKS, INC.

Chicago, Illinois

© 2011 Lyceum Books, Inc.

Published by

LYCEUM BOOKS, INC.
5758 S. Blackstone Ave.
Chicago, Illinois 60637
773+643-1903 (Fax)
773+643-1902 (Phone)
lyceum@lyceumbooks.com
http://www.lyceumbooks.com

6 5 4 3 2 1   10 11 12 13 14

ISBN 978-1-933478-65-4

Printed in the United States of America.

Library of Congress Cataloging-in-Publication Data

Miller, Joyce A.
  Improving performance in service organizations : how to implement a lean transformation / Joyce A. Miller, Tatiana Bogatova, Bruce Carnohan.
    p. cm.
  Includes bibliographical references and index.
  ISBN 978-1-933478-65-4 (pbk. : alk. paper)
 1. Service industries. 2. Organizational effectiveness. 3. Value added.
4. Corporate culture. I. Bogatova, Tatiana. II. Carnohan, Bruce. III. Title.
  HD9980.5.M547   2011
  658.4'01—dc22

                                                          2010021535

Contents

Illustrations

FORMS

BOXES

TABLES

Preface

As a sociologist with expertise in the evaluation of and quality improvement within social programs, I have an extensive history of applied research in a variety of organizations within the service sector. While these evaluations generally have been funded via government contracts, the programs evaluated provided a variety of services in the fields of education, health care, gerontology, substance abuse prevention, juvenile justice, and workforce development.

My approach in evaluating the programs operated by service organizations has generally included both outcome and process assessments, using mixed methods of quantitative and qualitative data collection and analysis tools. Furthermore, I have utilized participatory models of evaluation and focused on building the capacity of program stakeholders to plan for and complete ongoing evaluation with the intent that results of evaluation be used to make program improvements. This is the essence of organizational learning and continuous quality improvement.

However, the conceptual frameworks and methodological tools that I used to evaluate programs generally assessed whether a program was accomplishing its goals (i.e., its outcomes, inclusive of short- and long-term goals and objectives) and the quantity and quality of its program implementation (i.e., if a program provided the services as intended and if clients were satisfied with the services received, as an indicator of quality). As such, the focus was not on operational efficiency and the extent to which program operations were streamlined and resources were not wasted in an effort to accomplish program goals.

As a result of not addressing operational efficiency and resource use, programs could be effective and successful in doing what they were designed to do, yet be consuming scarce resources (e.g., person hours and funds allocated for physical and other material resources) that limited their ability to provide needed services to their client groups. Indeed, it is often the claim of social service organizations that they are underfunded and, therefore, cannot meet the demand for their services or cannot provide the level of intensity of their services, an intensity required to achieve the outcomes or have the long-term impact of their program efforts.

Having been introduced to lean and Six Sigma concepts and methodological tools in 2005, I realized that current evaluation models neglected to assess an important element of program implementation (i.e., operational efficiency) and its impact on effectiveness. Thus, evaluation as a field can benefit from a lean perspective and its use to comprehensively evaluate programs, in terms of both efficiency and effectiveness.

How did I come to this conclusion? It resulted from an organizational crisis of my own. In 2005, my company, KeyStone Research Corporation (KSRC), faced the loss of the only contract that it held. While it is obvious to most businesspersons that they should never "put all their eggs in one basket," through historical circumstances, I faced that situation. Over several years, KSRC was awarded a series of smaller evaluation contracts, but the financial stability of the company was based on the long-standing contract with a state agency to design as well as administer a statewide early care and education professional development system. It was a system that evolved over time under my leadership and grew from a small annual contract of $85,000 in 1988 to one that was worth over $13,000,000 in 2005.

With one year to wind down and transfer this contract to a new agency, I faced a severe crisis that jeopardized the survival of my business. Needless to say, it was a "wake-up call" that forced me to take a much closer look at my financial situation and core competencies and to determine how I would be able to generate additional sources of revenue that would keep my business solvent. It was at this point that I contracted the services of Bruce Carnohan, a lean and Six Sigma consultant, to assist me in the analysis of my operations and determination of cost-cutting measures I could put in place ASAP. The initial returns on this investment included cutting my overhead costs in half within one year.

As a result of my interaction with Bruce and observation of his approach to analyzing my operations, I had what might be considered an "eureka" moment. The lean and Six Sigma concepts were new to me, but I could see their applicability to other service organizations, particularly the type I had operated and evaluated over the years. Realizing the power of this process improvement methodology, I included a lean process analysis component in an evaluation project that KSRC was awarded in late 2005. The use of lean concepts and methods under this contract met with such success that the state funding agency amended our contract to include the analysis of additional processes within the existing program being evaluated, as well as processes within another division in the state agency.

As a result of this work and our presentation at professional conferences, we were offered other opportunities to build the capacity of other service organizations to understand and use lean concepts and methods in the analysis of their own operations and to determine ways to make significant improvements in their overall performance. The power of lean thinking was recognized by these organizations, which led me, along with Bruce Carnohan and Tania Bogatova, to move forward in writing this book. We are confident it will provide a useful guide to other service organizations in their effort to improve their performance through an application of lean concepts and methods.

While this book provides an introduction to lean thinking and its application to organizations in the service sector, it is not intended to be a comprehensive overview of all lean concepts and tools that are frequently used in the world of manufacturing. Nor does this book present Six Sigma concepts and methods and their application to service organizations. The key concepts and methods presented in this book were selected because of their immediate relevance to the work environment and types of processes frequently found in service organizations. Furthermore, this book sets the stage for introducing lean concepts and methods to service organizations by embedding this approach to process improvement within the broader context of organizational learning and continuous quality improvement (CQI). This broader context consists of conceptual frameworks and methodologies that are relevant to and generally understood by those working within service organizations; they are frequently used by evaluation researchers, applied social scientists, and management consultants who focus on leadership development and organizational change.

AUDIENCE

As discussed, this book offers a framework and set of tools adapted from lean philosophy and applied to service organizations, which provides a means to substantially improve performance. The concepts and methods offer an approach to analyze work processes and track performance over time, with a focus on how to continuously eliminate waste, engage in activities that are value-added, improve the overall efficiency of operations, and increase organizational effectiveness, ultimately.

The audience for this book is two-fold. First, the leadership, supervisory, and line staff within service organizations, such as government, education, health care, and other social services, will find the "how to" approach of this book instrumental in their development of knowledge and skills relevant to performance excellence. As written, this book includes numerous case examples, along with exercises to put into practice the learning that occurs as a result of reading the book chapters. The knowledge and skills acquired are relevant to the design phase of service programs, as well as the implementation and operational phases of programs where continuous learning and improvements are made.

Leadership at all levels within an organization will not only gain an understanding of the importance of measuring and tracking performance, but also will acquire the means to ensure that the processes embedded within service programs are streamlined, efficient, and effective in accomplishing their goals, with client needs as the driving force. Furthermore, given the focus on organizational learning, this book builds the capacity of readers to do more than give lip service to the notion that organizations should be continuously transforming, improving the quality of their services, and moving toward perfection of their operations. This book offers a tool kit to eliminate wasteful activities within processes and prevent unacceptable results, which are a function of process design and operation. In addition, management staff within service organizations will find this book useful in identifying the role for internal versus external "champions" to facilitate lean transformations and development of a culture of organizational learning.

The second audience for this book includes lean educators and facilitators (whether internal or external to an organization) who provide their expertise to service organizations undergoing a lean transformation. This book can be used by lean educators and facilitators to engage the leadership and line staff within a service organization and build their capacity for lean thinking. As articulated in this book, lean thinking must become ingrained within the culture and climate of an organization for the full potential of the application of lean concepts and methods to be reached. Finally, the acquisition of the knowledge and skills comes from an educational process that couples theory, methodology, and practice. The design of this book, with its grounding in lean philosophy as well as practical application, offers readers with more than just a "toolkit" to apply when faced with operational issues. The real power of lean thinking is that it offers a systematic way of learning by doing, recognizes that learning never stops, and reinforces the need to continuously change as an organization strives for perfection.

PLAN OF BOOK

This book has been organized in such a way to provide the conceptual knowledge as the foundation for all subsequent learning related to the development of specific skills and the use of tools to implement a lean transformation within an organization. Part I of the book includes two chapters that introduce the reader to the conceptual frameworks that are instrumental in establishing the importance of performance assessment and the improvement of processes within service organizations. Part II of the book presents two chapters that offer the basic concepts of lean philosophy and show how they can be adapted to service organizations. Part III of the book, three chapters, moves into the use of lean tools and methods with very specific applications to processes that are likely to be present within a service organization. The two chapters in Part IV turn to a discussion of the factors that can either enhance or inhibit an organization's ability to sustain process improvements over time, with specific cases and the lessons learned from the experience of organizations that implemented lean transformations.

In addition to the summary overviews for Part I, II, III, and IV of the book, the final section within each part, entitled "From Knowledge to Practice," offers specific exercises that can be used to further practice the application of knowledge and to assess the level of learning.

Appendices are provided at the end of Chapter 5 and Chapter 7, and at the end of the book. These appendices provide a number of tools that can be used for creating a logic model, diagnosing the extent of waste within a process, measuring and tracking performance data associated with a process; they also offer a checklist for doing a 5S audit and examples of an action planning tool and A3 report.

Finally, the end of the book includes a glossary along with a resource list that provides a variety of useful websites, software, and professional or trade associations that are relevant to the topics of lean, process improvement, performance assessment, continuous quality improvement, organizational learning, organizational change, and leadership.

ABOUT THE AUTHORS

Joyce Ann Miller, PhD

As a sociologist by training and education, I specialized in evaluation and social policy, graduating from Kent State University in 1981. In 1980, I established a private, woman-owned research and consulting organization, KeyStone Research Corporation (KSRC). In addition, between 1978 and 1999 I was professor of sociology at Villa Maria College and Gannon University and held the position of Associate Provost at Gannon University between 1995 and 1997.

My work has spanned several decades and includes rich experiences in a variety of "worlds." As the president and founder of KSRC, I am a small business owner and entrepreneur with an understanding of the business world. Through my work at KSRC, I had opportunities to develop, implement, and evaluate programs offered through service organizations, many of which were government funded. This experience provided me with greater insight as to the operations of government entities and their requirements for performance assessments and mandated accountability on the part of the organizations receiving government funding.

As an academic in higher education, and a sociologist by profession, I have been involved in numerous professional organizations, and have held many leadership positions. In a number of instances, I played an instrumental role in the development and implementation these professional associations or specific programs that were established to better serve their professional constituency. These experiences further expanded my knowledge and understanding of systems and processes, which must be designed to achieve organizational missions and goals.

Finally, while involved in the business world and academic or professional world, I worked within service organizations, primarily government and nonprofits that provide services to particular target groups, which are often seen as "clients" rather than "customers." Indeed, the service and public sector is often criticized because of its lack of understanding of business processes and inability to establish systems that are streamlined and that excel with respect to operational efficiency. Hence, the service sector is more than ready for an introduction to lean thinking.

Tania Bogatova, MBA

Tania has worked with executives, individuals, teams, and organizations to achieve sustained, accountable success for almost ten years. A capacity-building expert in the areas of program evaluation, development, and improvement, Tania holds an equivalent of Masters in Business Management and Economics from Sochi State University in Russia and Masters of Business Administration from Gannon University. She is currently completing a doctoral degree in Organizational Learning and Leadership from Gannon University. Tania has been a member of the organizational consulting and research team at KeyStone Research Corporations since 2003. Her work at KSRC has been varied, with a focus on evaluation research, as well as on the application of lean methodology for our service clients. She is very skillful in helping organizations successfully manage change—through training, coaching, and mentoring.

Bruce Carnohan

Bruce, as well, has worked with executives, teams, groups, individuals, and organizations to achieve exceptional results in the improvement of diverse business processes. Originally educated and seasoned in Europe, Bruce moved to the United States in 1994 through an international transfer with a major industry player in the dye and pigmentation fields for foods, pharmaceutical, and cosmetic coloring market sectors. Bruce has successful, hands-on process and quality improvement project management experience in more than one hundred different types of businesses, state agencies, and nonprofit organizations. His craft, methodology and know-how has been shared with corporations and clients for forty years and he has a somewhat unique way in looking at processes and systems that require redesigning; thus turning current unacceptable results and process waste into processes that are designed to deliver exceptional results. Bruce is passionate about helping organizations design the very best processes using the resources available to them. He achieves a great deal of satisfaction when he reflects on the "before and after" results achieved through the implementation methods and approaches shared in the contents of this book.

ACKNOWLEDGMENTS

Writing a book involves an intensive effort of having the experience about which one can write, conducting additional literature research and reviews, writing content, revising chapters, verifying references, getting permissions for reprinting material, and dealing with all the technical challenges of modern word processing and manuscript preparation (which is great when it works, but extremely frustrating when it doesn't). While one person can handle all of this, a team effort is definitely a better approach. Therefore, I want to extend my thanks to those individuals who were instrumental in bringing this book to fruition.

First, I must thank my colleagues who contributed to this effort. Bruce Carnohan was the catalyst for introducing me to lean thinking, as described above. His expertise in applying lean in a manufacturing environment and subsequently in my organization opened my eyes to a new way of assessing organizational performance and improving operational processes. His intellectual contributions provided the basis for applying lean thinking in the service world, where I am most familiar. He also provided additional input as to the content of the chapters and their organization.

Tatiana Bogatova, as the other contributor, was instrumental in overseeing the data collection and reporting activities for those sites where we facilitated lean implementations; supervising our intern, Jeffrey Hartman from Mercyhurst College in Erie, PA, who organized material from our literature review; seeking the required permissions; and preparing many of the charts and graphs, and gathering the details for the case examples in Chapter 9. Her attention to detail and never-ending enthusiasm and energy to work long hours never ceases to amaze me.

Another member of our office team, Vince Jarzynka, assisted in dealing with the technology challenges that always seem to happen when those pesky gremlins in cyber-space decide to play havoc. His patience and ability to find root causes for our technology glitches was greatly appreciated. Our graphic designer, Ann St. George Simpson, also assisted in the development of a number of our graphic items throughout the book and we thank her for her contributions.

The most important acknowledgment and thanks must be extended to those agencies and organizations that are highlighted as examples throughout the book, especially in our case studies described in Chapter 9. A number of these organizations served as our testing ground for applying lean concepts and methods in a service environment. There are numerous individuals from the organizations that were involved as leaders, staff members, and other stakeholders. While there are too many to mention by name, there are a few that deserve special recognition. At the Arkansas Department of Human Services, Division of Child Care and Early Childhood Education (DCCECE), we extend our thanks to Tonya Russell, Director; Donna Alliston, Professional Development Coordinator; and Kathy Stegal, Program Administrator. They provided us with the opportunity to use our lean methodology in an analysis of a number of their value streams within their early care and education professional development system. They also had the insight as to the power of lean thinking and extended our work to the Child Care Assistance program in an analysis of its subsidy and voucher application process.

Thanks are also extended to the South Carolina Child Care Center for Career Development (SC CCCCD), where we facilitated the mapping and analysis of their T.E.A.C.H. Early Childhood® Project and the Training Program for early care and educational professionals. Millie McDonald, State Director, was remarkable in that she fully committed to the implementation of lean in CCCCD and subsequently provided the leadership to sustain lean thinking in all CCCCD's operations. The other staff at CCCCD that were instrumental in taking lean concepts and methods and applying them in their programs included Debra Nodine, T.E.A.C.H. Coordinator and Training Coordinators, Ann Pfeiffer, Donna Davies, and Rebecca Dixon. Indeed, the SC CCCCD, is the "poster child" for the exceptional results that can be achieved through an application of lean thinking. Furthermore, SC CCCCD provides a model for leadership involvement and the development of a culture of organizational learning. Kudos to everyone there and may they have continued success in the operation of their programs.

In addition, we want to thank those individuals that facilitated our development of the two additional case studies described in Chapter 9. Tom Baumann, Director, Minnesota Office of Continuous Improvement, and Alisha Cowell, a member of his Enterprise Lean Program staff, were able to fill in the gaps and add insight into Minnesota's lean program. At the Fox Valley Technical College, we want to thank Cynthia Wetzel, Dirk Kagerbauer, and Barbara Kieffer, who were instrumental in the development of FVTC's Lean Performance Center and the lean transformation case presented in Chapter 9. Through our discussion with them at the 2009 Lean Educator Conference in Minneapolis, Minnesota, we believe their work provides an important example of how lean can work within an educational setting.

Finally, we extend our thanks to David Follmer, Nina Nguyen, Reese-Anna Baker, Siobhan Drummond, and Alison Hope at Lyceum Books for their guidance as we traveled through the long journey of completing this book. Their patience is greatly appreciated.

Joyce Miller

Introduction to Performance Improvement in Service Organizations

Improving performance should be a priority within your organization. After all, do you not want to say that your organization provides quality services or products to your clients? Unfortunately, determining how well your organization is performing, understanding the factors behind performance issues, and knowing how to improve your performance are not simple tasks. Moreover, there can be many obstacles or challenges within your organization that inhibit your ability to change and make improvements that will impact performance levels.

This book provides you with a framework and set of tools adapted from lean philosophy and applied to service organizations that can be used to improve your performance. The concepts and methods offer an approach for analyzing work processes and tracking performance over time, with a focus on how to continuously eliminate waste, engage in activities that are value-added, improve the overall efficiency of operations, and ultimately increase organizational effectiveness.

Part I of this book provides an overview of performance improvement concepts and methodologies and illustrates how lean transformations can be implemented in service organizations. Specifically, Chapter 1 sets the scene through its discussion of the increased focus on accountability and performance assessment that has been promulgated by the federal government over the past few decades. Furthermore, this chapter offers a preview of lean philosophy and how it must be adapted to apply to service organizations. The chapter concludes by highlighting the fundamental principles underlying the conceptual framework and

methodological tools that are presented in this book, which offer a means to improve performance though an analysis of processes and implementation of improvements that will produce exceptional results with respect to operational efficiency and effectiveness.

Chapter 2 establishes the context for improving organizational performance by taking a closer look at organizational systems, the drivers for performance-based accountability and program evaluation, and the quality movement that has permeated business and industry over the past several decades, including the service sector. Specifically, this chapter provides an overview of the open systems model for understanding organizational behavior, which is a good fit with the lean approach to process improvement articulated in this book. This chapter also examines the historical drivers for performance-based accountability, and then reviews a number of approaches to program evaluation, highlighting those that have significant relevance for process improvement efforts. Finally, we present a number of quality improvement models relevant to lean philosophy. We discuss the work of W. Edwards Deming, given his significant impact on the development of lean philosophy both in Japan and in the United States. This chapter reviews the Malcolm Baldrige National Quality Award Program, as it signaled the entry of the United States government into the quality movement. This chapter concludes with a discussion of two other approaches to process improvement—Six Sigma and the Theory of Constraints—since they are "cousins" of lean approaches to transforming businesses into more efficient and effective operations.

CHAPTER 1

Improving Performance Through a Lean Transformation

> *We ought to be interested in the future, for that is where we are going to spend the rest of our lives.*
>
> — Mark Twain

INTRODUCTION: SETTING THE SCENE

With the increased demand for accountability and performance assessment within government and other funding organizations, along with the influence of quality movements within business and industry, there has been a corresponding movement within the service sector (particularly health-care institutions, along with service businesses within the financial and insurance industry) to embrace a number of the methodologies used to improve their operations management and the quality of services delivered. Their focus has been reducing variation, improving turnaround times for the delivery of services, reducing the inconsistency and unpredictability in service interactions, providing defect-free services in a reliable manner, and designing systems to improve customer or client satisfaction, to name a few (Metters & Marucheck, 2007).

Along with this increased focus on accountability and performance assessment, many organizations in the service sector are facing an environment of rising costs and competing agendas for a limited pool of resources. This is particularly evident when organizations must cope with a growing demand for their services yet lack sufficient resources to expand operations, or when they are confronted with fewer resources to provide the same level of service. They must do more with the same or do more with less. Given this situation, service organizations must find solutions for their operational challenges if they are to maintain a level of effectiveness in meeting the needs of the clients they serve. Having established delivery systems that are effective, they must analyze the efficiency of their processes, since there may be a number of wasteful activities (WAs) that use up resources in a nonproductive way. Hence, it is imperative for service organizations to find avenues to improve the efficiency of their operations in ways that support and enhance their effectiveness in achieving organizational or programmatic goals.

This book offers a conceptual framework and how-to methodology that your service organization can utilize to improve performance. It offers a means to identify opportunities for improvements within your organization's operational and business processes that will produce positive results with respect to organizational effectiveness and efficiency. The model presented herein is an adaptation of lean philosophy, which has been used in manufacturing for several decades. It is an innovative use of some of the key concepts and methods of lean thinking within a different economic sector—service organizations and agencies.

In general, service organizations have lagged behind manufacturing and other businesses when it comes to operations management and improvement of processes (Hanna, 2007). It is routine for industrial engineers working in manufacturing environments to develop and maintain process analysis diagrams and blueprints for their automated, assembly line processes. Manufacturing offers opportunities to delineate a single, well-specified, standardized process to produce a product within a highly centralized and controlled environment. It is easier in a manufacturing environment than it is in a service environment to formulate a problem with a single objective (e.g., cost minimization) and to identify the conditions under which this objective can be achieved.

In service organizations, diagrams or blueprints for the delivery of services have not been developed routinely. Service organizations have resisted describing their services as processes, particularly as processes that are standardized and applied in the same way with all clients (Metters & Marucheck, 2007). Since client participation in the service interaction is a key difference between manufacturing and services, service system design has generally allowed a greater degree of variation, giving it features that are more customized and personalized. In addition, within service organizations there may be a number of processes that are not replicable or that occur so seldom that it is not worthwhile to apply techniques used to map out and analyze their efficiency and effectiveness. Regardless, an essential first step to improve routine processes is to develop detailed process maps and gather performance data associated with the process.

> *Maintaining a complicated life is a great way to avoid changing it.*
> — Elaine St. James

While not all operations management and process improvement ideas from manufacturing translate from the factory floor to the service organization, there are a number of improvement concepts and methods that can be adapted for a service environment. As in business and industry, it is important for your service organization to analyze the current design of your work processes and for you to learn how to improve those processes.

As an illustration of this compelling need, Example 1.1 describes a problematic situation facing a state government economic development agency. This agency's mission is

> to empower businesses and communities to invest, succeed, and thrive in an environment that affords a superior quality of life and increases opportunities for economic prosperity for the state.

There are a number of challenges associated with the effectiveness and efficiency of this agency's existing processes and systems. Specifically, the staff complement dedicated to the monitoring and compliance functions has not increased substantially over the years, yet the agency is faced with a rapidly expanding array of grant programs, grant and loan applications, and contracts awarded, along with an increase in accountability requirements. Between 2001 and 2008, the agency went from 4,500 to 7,500 contracts for awarded grants, representing a 66 percent growth. This is a situation where an organization has to do more with the same, which, without necessary changes to processes, compromises the agency's effectiveness in accomplishing its economic development goals. Furthermore, based on the list of concerns provided in Example 1.1, the process variation that exists across programs is a form of waste, in that it leads to inconsistency across grant programs with respect to monitoring and accountability, which is an unacceptable result (UR) for this agency. Without standardized or defined workflow processes in place across program areas, staff conduct their work differently and do not treat grantees equitably. This example represents a clear case where process analysis and redesign are needed to standardize the monitoring process and eliminate activities that are not value-added so that the agency can move toward a more-effective and more-efficient operation.

EXAMPLE 1.1. Government Economic Development Agency: Doing More with the Same

Agency	State Community and Economic Development
Mission	To empower businesses and communities to invest, succeed, and thrive in an environment that affords a superior quality of life and increases opportunities for economic prosperity for the state.
Problem Statement	• Administers more than eighty grant programs and five operating appropriations with grant allocations. • Total number of state and federal programs administered increased by twenty-five since 2001. • Total number of grant or loan contracts awarded annually increased from 4,500 to 7,500 since 2001. • Increasing demand for accountability in program administration. • Total staff complement dedicated to monitoring and compliance functions has not increased substantially since 2001. • Improved technology has improved ability to handle increased workload. • Operational and systemic challenges to monitoring, compliance, enforcement, and reporting capacities have surfaced.
Current State of Operational Processes	• Some program area staff positions can exercise broad, independent discretion in contracting activities and in contract management processes. • Some program area staff can make independent decisions to close out grant contracts without the timely knowledge of or engagement with staff in the office of audits and compliance. • There are several contract close procedure manuals in circulation within distinct program areas and offices. • The "flagging" of a noncompliant grantee (which, until corrective actions is taken, makes said grantee ineligible for additional or future grant awards) could go unreported to other program areas. • The protocol to "flag" grantees is not imposed uniformly in all programs. • Audit findings for which the appropriate corrective measures may have been identified and implemented are not adequately or uniformly tracked or reported in all program areas. • Program files can be misplaced or be incomplete and different program areas employ different protocols regarding file content, file management, or file archiving regimen. • Processes associated with the preparation, review, edit or update, and approval of program guidelines for final publications to the public are not uniform in all program areas.
Areas of Waste	• Process variation • Information deficits • Errors

THE SERVICE SECTOR AND LEAN PHILOSOPHY

What Is the Service Sector?

The service sector is broadly defined as that division of the economy that is service-producing as opposed to goods-producing. This sector includes a variety of industries that deliver services as well as material items (sometimes referred to as products) rather than manufacture products. The human element in service transactions is prominent, thereby introducing greater variability into the implementation of processes than what is found in machine or assembly line processes.

For our purposes, the focus of our application of lean philosophy to the service sector includes those industries that are categorized as public agencies and private organizations, the majority of which are nonprofit, although some may be for-profit entities. This includes organizations such as government agencies, health-care organizations, educational institutions, and social service or human service agencies. For the sake of simplicity and ease in communication, we will refer to the variety of service industries that are our focus as social service organizations or simply service organizations.

From our perspective, these service organizations are often described in terms of a logical sequence of inputs (various types of financial, human, and material resources) and activities that have been implemented to deliver a set of services or other tangible products or materials to a target group in order to achieve specific outcome(s) or result(s) for that group. The individuals within the target group are clients, and outcomes generally consist of changes in knowledge, attitudes, or behavior, which are changes that are associated with the goals of the organization based on client needs.

We do recognize that there are other types of service industries, particularly those classified as service businesses, that are providing services to a target group rather than producing a product on an assembly line. However, there are differences in perspective and terminology or use of language within service organizations versus service businesses. Table 1.1 shows some of these differences, although, over the years, there has been a cross-fertilization of concepts and use of language because of the trend in which

TABLE 1.1. Differences in Terminology: Social Service Organizations vs. Service Businesses

Social Services	Service Businesses
Organizational outputs	Units of service delivered
Clients (and specific labels for certain industries, e.g., students in education, patients in health care, etc.)	Customers (and specific labels for certain industries, e.g., shoppers in retail, buyers and sellers in real estate, etc.)
Client outcomes or results	Profit, cash flow, results
Client satisfaction	Customer satisfaction, customer loyalty
Vision, mission, goals, and objectives	Vision, mission, targets, benchmarks
Core values	Core values
Logic models, strategic plans	Strategic plans, business plans
Performance measures, outcome measures	Metrics, performance measures, dashboard, balanced scorecard, report card
Best practices, evidence-based practice, accreditation standards	Standards, licensing requirements, registration requirements
Accountability	Fiscal management, accountability
Evaluation, assessment	Auditing, assessment

service organizations are expected to be more business-like, and to have improvements including, but not limited to, improvements in operational efficiency, performance measurement and accountability, and customer satisfaction. Regardless, there are differences in orientation, with businesses primarily focused on profit and the bottom line, and with public agencies and social service organizations focused on providing goods and services that are considered universally desirable (e.g., national security) or good for a community as a whole to meet the needs of specific target groups (e.g., the aged).

With this in mind, throughout the book we will use the terminology and language most often used within public agencies and social services that are primarily nonprofit organizations. However, we do want to reiterate that the framework and methodology presented in this book is applicable to any process that involves the delivery of a service and transactional activities between the client or customer and the service provider. To that extent, this model also can be applied to administrative and office activities within manufacturing environments, but throughout this book, our focus and examples will come from public agencies and service organizations.

How Does Lean Philosophy Apply to Service Organizations?

To date, most of the theoretical and practical application of lean thinking has been in manufacturing (Womack and Jones, 2003). The essence of lean philosophy is to deliver the most value to customers while consuming the fewest resources. Value is defined from the vantage point of the customer (i.e., identify what the customer wants or needs). Once value has been defined in terms of customer wants or needs, waste must be eliminated with respect to time, expense, and material at every stage of the operating process to cost-efficiently solve the customer's problems and meet his or her needs so the organization can prosper (Womack & Jones, 1996, as cited in Marchwinski, Shook, & Schroeder, 2008).

While the historical roots of lean manufacturing are from the Toyota Production System (see Box 1.1), the terms "lean production" or "lean manufacturing" were introduced by researchers in the International Motor Vehicle Program at the Massachusetts Institute of Technology (Womack, Jones, & Roos, 1990). Their work articulates five key principles that together make up lean thinking (Barney & Kirby, 2004; Womack & Jones, 1996):

BOX 1.1. The Historical Roots of Lean Manufacturing

The historical roots of lean philosophy are in the Toyota Production System (TPS), which was pioneered by Eiji Toyoda and Taiichi Ohno at the Toyota Motor Company in Japan after World War II. This development was fueled by Henry Ford's development of flow production used to assemble the Model T. While the Model T production was revolutionary in its time (early 1900s), it had limitations with respect to continuity in process and variety of offerings. In a response to changing consumer demands for more variety and Japan's need to rebuild its industry after World War II, Toyoda developed the Toyota Production System where the focus was shifted from individual machines and their utilization to the flow of the product through the total process (Lean Enterprise Institute, 2008). The TPS is based on the principles of providing more choice to consumers, more decision-making involvement for workers, and more-efficient productivity to companies.

Today, TPS represents a revolutionary way of doing business that has expanded beyond the auto manufacturing industry. This system is being used in all types of manufacturing systems, administrative and office systems, and service environments, although the applications to nonmanufacturing environments do not clearly show how many of the concepts and tools applied on the factory floor are relevant in other types of business environments. It is with this challenge in mind that an application of lean concepts and methods to service organizations has been developed, as presented in this book.

1. Specify the value desired by the customer.
2. Identify the value stream for each product providing that value, and challenge all the wasted steps (generally nine out of ten) currently necessary to provide it.
3. Make the product flow continuously through the remaining, value-added steps (i.e., without interruption and wasted time).
4. Introduce pull between all steps where continuous flow is possible (i.e., a system that ensures production is tied closely to demand, so that no products are built until there is demand for them, and inputs are supplied at the appropriate time).
5. Manage toward perfection so that the number of steps and the amount of time and information needed to serve the customer continually falls (i.e., create a culture where everyone is dedicated to continuous improvement).

> *Be like a duck.*
> *Calm on the surface,*
> *but paddling like the*
> *dickens underneath.*
>
> — Michael Caine

While these five essential principles, along with others that have evolved from them, characterize lean manufacturing, there are three important principles that can be applied to service organizations, as described below. They are (1) focus on value stream, (2) standardization of jobs, and (3) worker empowerment (Barney & Kirby, 2004).

Focus on Value Stream. A value stream includes all the steps from beginning to end required to deliver a service or produce a product for customers. These steps may be both value-creating and non-value-creating from the perspective of the customer. Paying attention to the value stream has these advantages (Barney & Kirby, 2004):

1. Instilling in workers and managers a broad understanding of the process, the desired endpoint, and the role of each particular job along the way, which helps to foster the concept of jidoka (i.e., don't pass along problems to others).
2. Encouraging workers and managers to feel responsible not only for their particular task, but also for the final service or product.
3. Removing the black box between inputs and outputs, which allows problems to be traced to their root causes and dealt with appropriately.
4. Dealing with issues of service or product quality immediately and effectively because attention to the pathways of the process allows quality control to be built into processes rather than inspected at the end.

Standardization of Jobs. Standardization involves having work highly specified as to content, timing, sequence, and outcome. In lean systems, standardization minimizes variation and chance as much as possible in all process activities, thus helping to eliminate waste and ensuring that value flows through the process (Spear & Bowen, 1999). Through standardization, workers are better able to investigate their own work processes, similar to what a scientist would do using hypothesis testing and the scientific method to learn about what methods work and what do not. Therefore, standardization (i.e., control in the language of scientists) is a precondition to learning, which is key to continuous improvement, system learning, and organizational improvement (Adler, 1993). For example, a standardized work process that does not produce the results as intended or does not have workers with the skills to perform as required, establishes experiential knowledge that workers and managers can use to isolate the problems and determine solutions (Barney & Kirby, 2004).

Worker Empowerment. Worker empowerment is a critical factor in the development of a culture of continuous improvement since workers are the implementers of work processes and have knowledge about what does and does not work. The managers and other professionals that design processes based on theoretical models are best used as teachers and resource guides. In such an environment, managers are a source of support and expertise, and do not exercise their authority through command and control.

Empowerment of workers requires that they be trained in problem solving and work analysis techniques so they have the skills required to continuously find ways of making their work go more smoothly (Womack et al., 1990). Furthermore, they must be given the responsibility to identify problems and redesign processes as needed. The advantages of empowering workers include the following (Barney & Kirby, 2004):

1. Work processes are designed by the individuals who are most familiar with them and their specific knowledge is integrated into the design process, which enhances their buy-in.

2. Worker buy-in increases their motivation and interest in their own improvement.

3. With responsibility for problem solving at the worker level, a mechanism for finding and solving problems early, while they are still small and localized, is established.

> *Ability is what you are capable of doing. Motivation determines what you do. Attitude determines how well you do it.*
>
> — Lou Holtz

What Are the Challenges in Applying Lean to Service Organizations?

Because lean's roots are within the manufacturing sector, the use of lean concepts and methods within service organizations comes with its challenges. Two critical challenges are (1) specifying what is of value to clients and (2) focusing on providing needed services and not on profit.

Specifying What Is of Value to Clients. Within service organizations, although the clients' needs should be the driver in the design and implementation of programs and processes that deliver services to them, there may be situations where client perceptions of their needs and satisfaction with services are not the key factors in analyzing processes to determine if they contribute to what is of value to the client. For service organizations, there may be different external as well as internal clients from whose perspective value is determined. For example, a childcare resource and referral agency's mission might be to improve the quality of early care and education by providing access to college education for poorly educated childcare staff. This may not be of value to those staff, however, many of whom are unaware of their educational needs and have only minimal educational requirements for their job. The professionals within the social service organization (i.e., internal clients) are the drivers of the system that creates value (improved knowledge and skill) for the childcare staff (i.e., external clients).

In addition to these primary internal and external clients, it is important to keep in mind that in social service organizations the client may include not only individuals, but also the system within which those individuals are stakeholders. Thus, in the professional development (PD) system for childcare staff, the staff are the recipients of PD opportunities (i.e., individuals as clients). Also, since the expected impact of PD is to improve the quality of early care and education, the facilities within which childcare staff work are also clients, since they need to have staff that can provide quality care. At another level, there is also a super client, which in this situation consists of the parents and children, because they are the ultimate beneficiaries of the improved quality of child care.

Focus on Providing Needed Services, Not on Profit. Undoubtedly, not all of the lean manufacturing ideas make sense within a service organization. In business and industry, the primary focus is on profit and how to improve operations to increase the bottom line. In many social service organizations, the focus is not on profit, and the notion of making a profit may even take on a negative connotation. A majority of public agencies and social or human service organizations focus on supplying services that are needed and considered good for the community as a whole or for a specific target group rather than on maximizing profit.

Regardless, whether or not profit is the driving force for your service organization, it makes sense for you to examine the design of operational processes to determine ways to eliminate those activities that are wasteful and add no value to the services you deliver. Your ability to use lean knowledge and its tools to ensure sound programmatic and fiscal management will contribute greatly to your overall success and ability to raise funds for your operation. In essence, your organization needs to find opportunities to optimize the use of time and money associated with your operational processes so that your scarce resources can be allocated to those activities that will produce the greatest value for the clients you serve. Ultimately, this increases your organization's accountability and fiscal responsibility. There are lessons you can learn by taking a look at the concepts and methods used within business and industry to maximize profits and establish a culture of process improvement.

Beyond these key differences with respect to organizational purpose, client or stakeholder groups, and profit motive, there are some similarities between the social service sector and the manufacturing or business sector. First, both social service organizations and business or industry want to be successful in what they are doing. They both have customers or clients that receive the product, information, or service being sold or delivered. Furthermore, the customer or client expects something to happen as a result of buying a product, obtaining information, or receiving a service. A customer that purchases a car expects that the car will run so that traveling from one point to another is an efficient and pleasant experience. A client that participates in an educational program expects to gain knowledge and be able to apply that knowledge appropriately, i.e., to put into practice what has been learned. In both cases, customer or client satisfaction is important, and so is customers or clients getting the results that they expect.

ADAPTING LEAN FOR YOUR SERVICE ORGANIZATION

Given the similarities as well as differences between the social service and manufacturing sectors, the question becomes, "How can you use a lean toolkit to implement a lean transformation within your social service organization?" Our purpose is to show you that there are core elements of lean philosophy that can be applied within your social service organization where there are processes in place to accomplish a goal.

As articulated in this book, lean thinking represents a method of problem solving, with its associated principles that guide thought processes and decision making in the effort to identify and implement solutions to problems. Moreover, our book does not simply supply you with a toolkit that you can apply without first addressing your organization's readiness to undergo a lean transformation. We introduce you to the factors that are essential in creating a culture and climate of excellence within your organization, which is the foundation of organizational learning and essential if you are to engage in a cycle of continuous improvement and innovation.

Ultimately, our goal is to build your understanding and capacity to identify and eliminate activities that you consider wasteful and not essential in delivering what is valued by your clients or in meeting your clients' needs. Specifically, you will be provided with a means to identify problems within processes, particularly those problems associated with WAs, and to design a future state of those processes with improved effectiveness and efficiency.

Overview of Proposed Framework and Methodology

The framework and methodology presented in this book provide you with the means to (1) analyze work processes with respect to value-added (VA) and non-value-added (NVA) activities, (2) determine the root cause of URs of processes, (3) identify opportunities for improvements, (4) specify the means for transforming the organization to implement the improvements, and (5) engage in a cycle of continual organizational learning.

The application of this framework and methodology within your organization will require you to implement a systematic, sequential, and cyclical set of steps, which are described in detail in Part II and Part III of this book. These steps include the following:

1. Gathering data via surveys, key informant interviews, observation, and document review to identify processes that are problematic to both internal and external clients that experience the processes. Specifically, the purpose of the data collection is to identify processes with WAs that lead to significant URs within your organization.
2. Establishing a core team of organizational staff and stakeholders to map the identified work processes.
3. Mapping the identified work processes (the current state).
4. Specifying key performance measures for the processes and organizational outcome measures.
5. Identifying opportunities for eliminating NVA and WAs within these processes.
6. Designing the improved and streamlined processes (the future state).
7. Developing action plans for implementing the process improvements.
8. Tracking performance measures over time.
9. Establishing mechanisms to assess results of process improvements.
10. Utilizing the knowledge for continuing the cycle of change, improvement, and innovation.

Fundamental Principles of Proposed Framework and Methodology

There are a number of fundamental principles that lay the foundation for our proposed model for improving your organization's performance through a lean transformation, as summarized in Table 1.2 and discussed below.

Organizational Vision, Mission, and Goals. This model uses your organization's vision, mission, and goals as a starting point since they create the basis for your existence and establish the context within which your organizational structure and operational processes are designed and implemented. The driving force for your service organization is to deliver services to targeted individuals or groups to meet identified needs and achieve intended outcomes for your organization's client base. Therefore, process improvement efforts are not about efficiencies only. It is possible to redesign processes so they are more efficient (e.g., they eliminate steps in a process and save time). However, if such efficiencies result in unintended outcomes and compromise the accomplishment of your organization's mission to meet the needs of your clients, then such process improvements will be counterproductive and should not be implemented.

TABLE 1.2. Fundamental Principles of Proposed Framework and Methodology for Implementing a Lean Transformation

ORGANIZATIONAL VISION, MISSION, AND GOALS

Your organization's vision, mission, and goals are a starting point for improving performance through a lean transformation since they create the basis for your existence and establish the context within which your organizational structure, operational processes, and specific programs are designed and implemented.

PROGRAM LOGIC MODEL

Your process improvement efforts cannot proceed without having a defined framework for understanding your organizational program or service under scrutiny. This is provided by a logic model that delineates the connection between your inputs, activities, outputs, outcomes, and impact.

TEAM APPROACH

Your efforts to improve organizational performance must utilize a team approach to analyze processes and develop improvements, since your operational processes have been designed and are being implemented by individuals at different levels within your organization.

STANDARD WORK PROCESSES

Your efforts to improve performance through a lean transformation will require the development of standardized work processes. Standardization increases the likelihood that similar results will occur when a process is completed by multiple staff or the same staff person repetitively over time.

CLIENT NEEDS AND WASTEFUL ACTIVITIES

The essence of this approach to improving performance through a lean transformation is to identify waste that is embedded within your processes and to take steps to eliminate the wasteful activities that do not meet the needs of your clients.

PERFORMANCE AND OUTCOME MEASURES

Your lean transformation will require tracking performance and outcome measures over time. As such, you must reach a consensus as to your performance and outcome measures, how they are operationalized, the frequency of their measurement, the method of reporting them, and their utilization for further process improvements.

Program Logic Model. If not already established, process improvement efforts cannot proceed without having in place a defined framework for understanding your organizational program or service that is under scrutiny. Having a logic model that delineates the connection between your inputs, activities, outputs, outcomes, and impact can provide such a framework, as depicted in Figure 1.1. Establishing such a framework prior to any process improvement effort helps to establish consensus among your staff and stakeholders about organizational or program purpose and strategy for accomplishing the goals associated with that purpose.[1]

Team Approach. This approach to process improvement utilizes a team approach to analyzing processes and developing improvements. Most likely, organizational or program operational processes have been designed and are being implemented by individuals at different levels within your organization. For example, you probably have management staff that are responsible for the overall design of processes, and you probably have line staff that carry out the day-to-day operations associated with the processes. Furthermore, you may have other stakeholders that interface with or are affected by your processes. Given this, it is important to incorporate the views of such individuals who view processes through their own lens and have different perspectives as to the operation of the processes. Therefore, when you map and analyze your processes to

---

[1] Creating a logic model for your organization or a specific program you operate is not something we will address in detail within this chapter. Instead, we have offered additional detail on the component parts of logic models and provided an example for you in Appendix A.

FIGURE 1.1. Logic Models

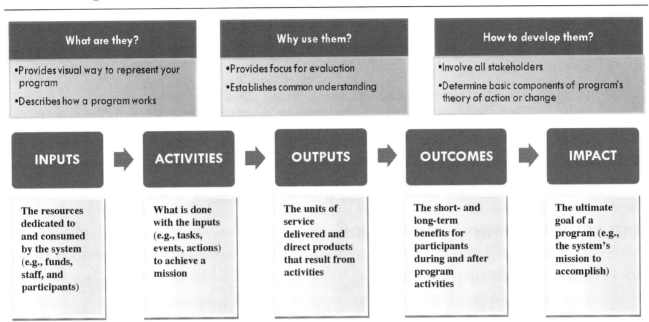

identify improvements to implement, it is important to establish a team of individuals representing the different perspectives. Furthermore, the establishment of representative teams will foster consensus building and collegiality, which are critical for creating an environment where continuous process improvement and innovation can be established and maintained.

Standard Work Processes. This approach to improving performance through a lean transformation emphasizes the importance of creating standardized work processes. Many of your organizational processes may not be completed by multiple staff or the same staff person repetitively over time, with the expectation that similar results occur. For those processes that are repetitive, it is important for you to have clearly defined steps that help to eliminate an unacceptable degree of process variation. Processes that vary in the way they are completed by the same person from one time to another, or that vary in the way in which different people carry out the same process, will lead to quality errors. And quality errors can lead to rework—another wasteful activity that is a function of the way in which a process is currently designed. Standardizing your work processes will add discipline to the work culture and ensure that your processes are consistently carried out over time, giving you predictable results. Also, having standard work processes will facilitate your training of new staff and provide you with a baseline for further improvements.

Client Needs and Wasteful Activities. The essence of this approach to improving performance through a lean transformation is to identify and take steps to eliminate WAs that are embedded within your processes. Waste can come in many different forms; what is considered waste to one person may not be waste to another. Therefore, it is necessary to have a basis for defining waste. As established in lean philosophy, the needs of your clients form the basis for these decisions. If a process step adds no value to the delivery of a service or product to meet your clients' needs, then that process step

13

becomes the focus for determining if it can and should be eliminated. There may be situations where your process steps may not add value from your clients' perspective, but are necessary due to other organizational, stakeholder, or regulatory requirements. In this situation, the process steps are considered required NVA activities.

Performance and Outcome Measures. This model requires the tracking of performance and outcome measures over time. You must reach a consensus as to your performance and outcome measures, how they are operationalized, the frequency of their measurement, the method of reporting them, and their utilization for further process improvements. This measurement activity is a critical part of the continuous quality improvement cycle involving stages of planning, implementing, evaluating, and improving, as further discussed in Chapter 2. If you carry out this cycle with integrity and if it is ongoing, you will foster organizational learning and facilitate decision making with respect to strategies that enhance your organizational success.

BENEFITS FOR YOUR SERVICE ORGANIZATION

In summary, taking these six fundamental principles together, our proposed model to improve your performance through a lean transformation will have a number of benefits for your organization. It offers you an innovative approach for analyzing your processes and identifying opportunities for improving your organization's performance. A lean transformation

1. promotes organizational learning throughout all levels of your organization;
2. builds the internal capacity of your team members to analyze the design of your work processes, identify improvements, and prioritize their implementation;
3. empowers your staff and other stakeholders to implement process improvements in the short term and continuously over time;
4. focuses organizational efforts to meet client needs by establishing what is VA from the perspective of your clients and stakeholders;
5. reduces process variation by standardizing your work processes;
6. improves your organization's efficiency and effectiveness by reducing WAs and eliminating URs due to the design of your processes; and
7. enhances the methods you use to assess processes and outcomes of program efforts on a continuous basis.

It is important to recognize that, in practice, using this approach to performance improvement through a lean transformation does not involve the use of a detached, outside expert who comes into your organization to analyze your work processes and recommend improvements; that is a top-down approach to organizational change and rarely works. While it may be necessary for you to use an outside expert to guide you through an application of this methodology when you initiate efforts to improve your performance through this approach, such an expert should build your capacity to apply this methodology on your own.

For the most part, this model for applying lean in-service organizations has been designed with the recognition that it is vital for you to create a culture of continuous improvement and innovation if your changes are to be implemented and sustained over time. It is essential that your organizational members and stakeholders recognize they have control over operational processes; as an organization, you get what you design, which may be positive or negative with respect to your organization's performance.

Chapter 7 further discusses the factors that contribute to the development of a culture of continuous improvement and innovation; in the meantime, it is important that you have an awareness of these factors before you begin your lean journey. As an organization, you must

1. have organizational leadership that fully supports and commits to changes;
2. identify your improvement team (at all levels of your program's operation, including clients and other stakeholders) and get all members' commitment to map and develop realistic action plans;
3. convince your staff that it is important for them to be open and willing to conduct work in a different way;
4. assure your staff that they can be honest and communicate the real processes without fear of retribution;
5. have your team provide the level of detail needed in process flow mapping;
6. determine whether a task is VA (i.e., from the perspective of your clients, staff, or funders);
7. access the needed resources for your change efforts;
8. find useful and easy-to-implement performance measures to track over time;
9. determine ways to sustain your process improvements and momentum of change over time; and
10. dispel the assumption that improving your work processes eliminates jobs, rather than frees up and better utilizes your organization's resources to accomplish your goals and objectives.

Establishing a Context for Improving Organizational Performance

---

CHAPTER 2 AT A GLANCE

OPEN SYSTEM MODEL OF ORGANIZATIONS
- Underlying Assumptions
- Key Elements
- Organizations as Systems and Logic Models
- Understanding Organizational Change
  - Continuous Change Cycle
  - Essential Elements of Change Management

DRIVERS FOR PERFORMANCE-BASED ACCOUNTABILITY AND PROGRAM EVALUATION
- Federal Legislation and Initiatives
- Program Evaluation in Service Organizations

- Approaches to Evaluation
- Limitations of Current Evaluation Approaches

QUALITY IMPROVEMENT MODELS
- W. Edwards Deming
- Malcolm Baldrige National Quality Award Program
- Six Sigma
- Theory of Constraints
- Quality Models and Improving Organizational Performance

To provide a context within which we can articulate the proposed model for improving organizational performance in service organizations, it is essential for you to have an understanding of (1) the basics of organizational systems, (2) the drivers for performance-based accountability and program evaluation, and (3) the quality movement that has permeated business and industry, including the service sector.

To that end, we begin with a discussion of the open system model of organizations. This model establishes the framework for conceptualizing the basic components of organizations and how these components work together in a functioning organization. Within this systems framework we provide a discussion of organizational change and the critical features of organizations that can either facilitate or inhibit the management of organizational change.

OPEN SYSTEM MODEL OF ORGANIZATIONS

At a macro level, all organizations have a structure and a set of processes that describe their operations. While organizational theories offer a number of ways to conceptualize both the static and the dynamic aspects of organizations, viewing organizations as systems is essential for our purposes because it offers a framework for understanding organizational change within the context of both internal organizational features and an organization's external environment, both of which impact organizational behavior.

Underlying Assumptions

The major components of an organization, based on the open system model, include the external environment, the organizational divisions or departments, the individuals within those divisions, and the reciprocal influences among the various organizational elements and the external environment. The underlying assumptions of the open system model and its various elements include the following (Harrison, 1994):

1. **External environment.** The social, political, and economic conditions external to an organization that influence the inputs into an organization, affect the reception of outputs from an organization's activities, and directly affect an organization's internal operations.

2. **Interrelationship of system elements.** All system elements and their subcomponent parts are interrelated and influence one another in a multidirectional fashion (rather than through simple linear relationships).

3. **Systems and subsystems.** Any element or part of an organization can be viewed as a system in and of itself.

4. **Feedback loop.** A feedback loop exists whereby the system outputs and outcomes are used as information that has a subsequent impact on system inputs over time, with continual change occurring in the organization.

5. **Organizational structure and processes.** The structure and processes within an organization are determined, in part, by the external environment and are influenced by the dynamics between and among organizational members.

6. **Organizational success.** An organization's success depends on its ability to adapt to its environment, to tie individual members to their roles and responsibilities, and to manage its operations over time.

> *It's never too late to become what you might have been.*
>
> — George Eliot

Key Elements

Figure 2.1 provides a graphic depiction of the key elements incorporated into an open system model of organizations. This model shows the internal elements of an organization, consisting of its structure and processes that define an organization's operations and day-to-day functioning. Since many service-based organizations have multiple programs that they operate, this model takes that complexity into account. An organization's

FIGURE 2.1. The Open System Model of Organizations

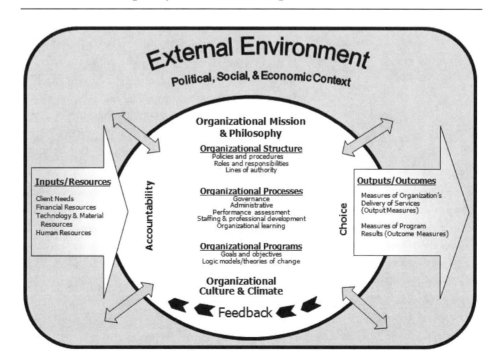

mission and philosophy serve as the driving force behind organizational activities and give purpose to its existence.

The system inputs consist of a variety of resources that an organization needs to operate its programs and fulfill its other administrative functions, such as human resources, financial resources, technology and equipment, knowledge, and legal authorizations. The outputs of an organization are the products and services delivered through the activities of the organizational members and the outcomes are the results of those activities for the organization's target group.

All these elements function within the context of an organization's prevailing norms, beliefs, and values (culture) and the current patterns of behavior, attitudes, and feelings that characterize life in the organization (climate), both of which inform individuals within the organization about acceptable and unacceptable behaviors, around which action is organized. Furthermore, the organizational system is dynamic, as depicted by the feedback arrow showing how information flowing from the outputs and outcomes has an impact on future inputs and resources.

Finally, the external environment consists of a task environment embedded within a general political, social, and economic environment that directly influences an organization's internal environment and operations. The task environment includes all the organizations and conditions that are directly related to an organization's purpose and operations. The components of the task environment have a more immediate impact on an organization. However, the task environment exists within a broader, more general social, political, and economic set of conditions that have a long-term, indirect impact on an organization and its task environment.

Organizations as Systems and Logic Models

While the open system model provides an abstract view of organizations with broad generalizations regarding organizational behavior, to have a better understanding of what organizations do and why they do it we need another level of detail. It is with this level of detail in mind that our approach to process improvement includes an essential first step, which is the development of a logic model to describe your organization's program or service under scrutiny. (See Appendix A for more details on the component parts of logic models.)

As discussed in Chapter 1, having a logic model delineating the connection between your inputs, activities, outputs, outcomes, and impact provides the framework for process improvement efforts, helps to establish consensus among your staff and stakeholders about the organizational or program purpose and the strategy for accomplishing the goals associated with that purpose. Without this framework, improving an organization's performance through process improvement efforts will lack focus and direction, particularly with respect to your organization's ultimate purpose, which is to serve your clients.

Understanding Organizational Change

Undoubtedly, there are a number of theoretical models and methodological approaches that help us understand how organizations are structured, how they design their processes, and how change occurs. For our purposes, it is essential that these models emphasize the importance of continuously improving organizational systems and using systematic approaches to understand what organizations are doing, how well they are doing it, and what changes need to be made to improve their performance. Both the open system model of organizations and logic models support this type of thinking about organizational behavior and management of change. The continuous change cycle can be depicted in terms of the phases discussed below.

Continuous Change Cycle. The continuous change cycle includes four basic phases, as shown in Figure 2.2:

1. Phase I: Planning for a change
2. Phase II: Implementing the planned changes
3. Phase III: Evaluating or assessing the results of the implementation
4. Phase IV: Taking action based on learning that occurred in the evaluating or assessing phase

In the quality control, lean, and Six Sigma literature, this cycle of continuous change has been called the PDCA cycle, representing the phases of Plan, Do, Check (or Study), and Act (Deming, 1986; Shewhart, 1939) or DMAIC, representing the phases of Define opportunity, Measure performance, Analyze opportunity, Improve performance, and Control performance (Motorola University, 2010).

While the labels attached to the phases may differ, this cycle has its roots in the scientific method: (1) stating a hypothesis about cause-and-effect relationships, (2) implementing an experiment to test the hypothesis (i.e., an intervention designed to bring about the expected change), (3) gathering the data to evaluate or assess the results of the experiment, then (4) modifying the intervention based on the knowledge gained as to whether or not the intervention worked.

FIGURE 2.2. Continuous Change Cycle

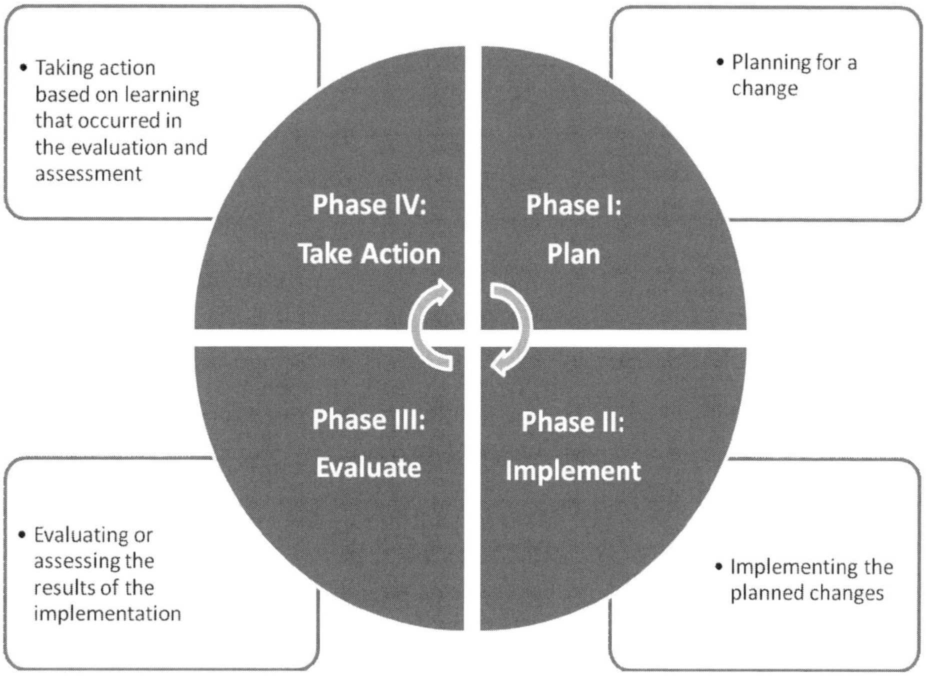

With respect to a continuous change cycle, if a planned change does not produce the desired results, then the cycle is repeated with a new plan. If the planned change works, then it may be that the change is applied in a wider arena, or that other changes are planned and the cycle begins again. In essence, change in organizations is never-ending; this cycle represents a way to design and implement changes in a logical and systematic fashion so that chaos does not rule the organization.

Essential Elements of Change Management. Beyond this basic premise that change and improvement are a continuous process, there are other essential elements that are instrumental for understanding organizations and the change process, including the following:

1. **Systems thinking:** Viewing organizations as complex systems with a structure and set of interrelated processes.
2. **Leadership:** Recognizing the role that leadership plays in the success (or lack thereof) of change efforts.
3. **Worker empowerment:** Emphasizing the need to secure front-line worker buy-in for creating a culture of change and quality improvement.
4. **Organizational goals:** Identifying the purpose of processes that guide the design and implementation of process activities.
5. **Performance measures:** Identifying and using metrics to have evidence of what works and what does not work.

All the models for improving performance through process improvements, whether used primarily in manufacturing or in service organizations, have something to offer with respect to how organizations or businesses gain an understanding of what they do, how well they do it, and the steps they need to take to ensure the achievement of their goals and objectives.

However, you need to recognize that there is no magic bullet that can be used by your organization to solve all your problems and move you toward excellence with respect to your operational efficiency and effectiveness. Change agents within your organization must have an understanding of the different approaches to improving performance and the tools available, since it will provide you with a knowledge base from which useful tools can be selected.

DRIVERS FOR PERFORMANCE-BASED ACCOUNTABILITY AND PROGRAM EVALUATION

As articulated in the open system model, the external environment has an impact on an organization and its activities. In light of that, we take a step back and review a number of historical developments to identify the roots of today's focus on accountability and performance-based assessment. These historical developments have affected business and industry, as well as service organizations, particularly public (government) and non-profit organizations.

Federal Legislation and Initiatives

While there were a number of seminal thinkers in the United States that influenced the quality movement in business and industry after World War II (e.g., see the discussion of the work of W. Edwards Deming below), the application of their quality improvement models had little impact on American businesses and service-based organizations until the 1960s. Beginning in the 1960s, there was a series of federal initiatives and

> *You must be the change you wish to see in the world.*
> — Mahatma Gandhi

pieces of legislation that served as a catalyst for what we see today as a significant focus on accountability and performance-based assessment.

In the 1960s and 1970s, coinciding with President Johnson's War on Poverty and the significant increases in the federal government's investment in educational and social programs, there was a growing concern about accountability to the American people and the use of federal money for such purposes. This resulted in a requirement for evaluation of programs funded through the Elementary and Secondary Education Act (ESEA) and of Medicare, Medicaid, and other social programs (Shadish, Cook, & Leviton, 1991). During this time, other performance-based management initiatives promulgated by presidents included the Planning-Programming-Budgeting System (PPPS) under President Johnson, the Management by Objectives under President Nixon, and Zero-Based Budgeting under President Carter (Radin, 2006).

In the 1980s, under President Reagan's mantra of less government involvement, funding for new social initiatives was cut and there were fewer requirements for an evaluation component of federally funded programs. "Government agencies were not required to report much beyond program inputs and outputs to justify their existence" (Carman, Fredericks, & Introcaso, 2008, p.8). However, at this same time, government agencies at the state and national levels, along with school districts, universities, and private companies, established internal evaluation units.

Like a pendulum swinging back and forth, in the late 1980s and 1990s there was a resurgence in the government's call to hold programs accountable and to track their performance. This was promulgated under the Clinton administration via the National Performance Review, which exemplified the pressure to reinvent government and streamline bureaucracy. It was important for organizations "to be lean, efficient, global, and more competitive" (Preskill & Russ-Eft, 2005, p. 4). With this renewed interest in accountability and performance assessment, legislation was passed. One of the most significant pieces of legislation ever passed in this area was the Government Performance and Results Act; this act is further described in Box 2.1, Government Performance and Results Act (GPRA).

While GPRA is federal legislation, its impact has had a domino effect down the line, with state and local governments being held accountable in similar ways because of the federal dollars that flow through block grants and other direct funding of programs and services. As discussed later in this chapter, the nonprofit world of funding, through foundations and agencies such as the United Way and the W. K. Kellogg Foundation, has recognized the value and importance of results-oriented performance planning, measurement, and reporting, and has used such data to make decisions about who and what programs to fund.

As a result of the governmental focus on accountability, there has been a considerable amount of activity within human service environments that rely on government and foundation funding to engage in strategic and performance planning and reporting. Programs are far more cognizant of and explicit about the outcomes they want to achieve (in the short and long terms), as well as the ways in which inputs and resources, activities, and outputs are linked as part of the process to achieve the stated outcomes. As previously discussed, the use of logic models to describe these linkages is now well established within the human services, since they enable program planners to think critically about what they are doing, to articulate their theory of change, and to logically plan their program and its evaluation (Bennett & Rockwell, 1995; Miller, Simeone & Carnevale, 2001; Rush & Ogborne, 1991).

BOX 2.1. Government Performance and Results Act (GPRA)

The Government Performance and Results Act (GPRA), referred to as a "results-based accountability system," was a key government performance-based management initiative in the 1990s (Office of Management and Budget, 2008; U.S. Department of Labor, 2006). From President Clinton's initiative to reinvent government so that it works better and costs less, to the Republican calls for devolution of services from the federal to state and local levels, this landmark legislation signaled another era where organizations were to be held accountable for the services they provided and the results they produced. This government action revived program evaluation and brought the measurement of outcomes and performance assessment to the forefront of public, as well as nonprofit, management (Carman et al., 2008).

Initially, GPRA grew out of a concern about the efficiency and effectiveness of public programs. GPRA was distinctive: it was a legislative mandate and tied budgetary decisions to performance results, although it did provide for considerable "start-up" time (Office of Management and Budget, 2004; Radin, 2006). It had three main requirements for federal agencies: (1) produce a strategic plan that includes organizational goals and objectives, (2) generate a performance plan that includes measurement and data on meeting objectives, and (3) compile a performance report that includes actual performance data (Bass & Lemmon, 1998). The purposes of GPRA are to

1. improve the confidence of the American people in the capability of the federal government, by systematically holding federal agencies accountable for achieving program results;
2. initiate program performance reform with a series of pilot projects in setting program goals, measuring program performance against those goals, and reporting publicly on progress;
3. improve federal program effectiveness and public accountability by promoting a new focus on results, service quality, and customer satisfaction;
4. help federal managers improve service delivery by requiring that they plan for meeting program objectives and by providing them with information about program results and service quality;
5. improve congressional decision making by providing more objective information on achieving statutory objectives and on the relative effectiveness and efficiency of federal programs and spending; and
6. improve internal management of the federal government (Office of Management and Budget, 2008).

As we enter the new millennium, government's focus on performance-based management and accountability has continued, as reflected by President Bush's implementation of the Program Assessment Rating Tool (PART). This tool is used by the Office of Management and Budget to score government programs in an effort to assess and improve program performance so the federal government can achieve better results. Furthermore, PART is used to make budget decisions based on results. In addition, there is a requirement in PART that programs (1) use only evidence-based practices to address social problems or issues, (2) measure their own performance, and (3) document that their efforts are having the intended impact.

To get what you want, STOP doing what isn't working.

— Dennis Weaver

Program Evaluation in Service Organizations

It is difficult to disentangle the development of the evaluation field and the federal initiatives and legislative actions described above. Without a doubt, the evaluation profession in the United States was significantly impacted by the government-funded evaluation initiatives of the early 1960s. Over time, evaluation evolved to what it is today as a result of government "reinvention, devolution, and performance-based management initiatives" (Carman et al., 2008, p. 6). As described in Box 2.2, The History of Evaluation, the growth and development of the field is linked closely with government initiatives. Today, evaluation as a profession is growing. There are a number of approaches to evaluation that have significant implications for process improvement efforts, as discussed below.

BOX 2.2. The History of Evaluation

Prior to the late 1950s and early 1960s, social science researchers focused their evaluation efforts on developing and implementing educational assessments. However, growing out of President Johnson's War on Poverty in the mid-1960s and the subsequent presidential performance-based management initiatives, the expectation was that evaluation would provide an understanding of the causes of social problems and the means to fix such problems. During this time, professional evaluation grew exponentially as a result of the federal requirements for the evaluation of programs.

During the next era (1970s and 1980s), concerns were voiced about the utility of evaluation findings and the methodological designs used in evaluation research (i.e., experimental and quasi-experimental designs). There were calls to ensure that evaluation had practical applications in terms of improving program effectiveness, enhancing organizational learning, and providing input for decision making, particularly with respect to resource allocation. With this growing emphasis within the public sector on measuring outcomes and being held accountable, there were calls for more-effective philanthropy coming from the private, nonprofit sector (Newcomer, 1997). Given this, a number of foundations created internal evaluation units and developed support for program evaluation activities (e.g., the work of the United Ways and the support they received from the Lilly Endowment, the W. K. Kellogg Foundation, and the Ewing Marion Kauffman Foundation) (Hendricks, Plantz, & Pritchard, 2008).

Since the 1990s, evaluation practice has evolved, with growing interest in different approaches to evaluation. There is a recognized need for participatory, empowerment, collaborative, and learning-oriented formative evaluations, rather than only summative or outcome evaluations that are conducted by lone evaluators who have the expertise and requisite skill to conduct evaluations. These alternative approaches to evaluation acknowledge that few evaluations are value-free and that most are politically charged. Use of these new approaches further recognizes the importance of having stakeholders participate in the development of evaluation questions, designs, and analytic techniques to enhance the utilization of findings.

In 2001, the reauthorization of ESEA, which became known as the No Child Left Behind (NCLB) Act, had substantial implications for the practice of evaluation. The NCLB Act is considered the most sweeping elementary and secondary educational reform since 1965; it has fostered a considerable amount of debate over its efficacy (Miller, 2002). Specifically, its requirements and position on accountability and performance assessment, which became a signature component of this act, have been fiercely debated. This act placed an increasing emphasis on evidence-based practice and the use of randomized control group designs (the gold standard for research on the effectiveness of interventions) to ensure that proven teaching methods are used as a means to guarantee positive outcomes for children. This governmental mandate with respect to the use of randomized control group designs led to the American Evaluation Association's response that the legislation manifests fundamental misunderstandings about (1) the types of studies capable of determining causality, (2) the methods capable of achieving scientific rigor, and (3) the types of studies that support policy and program decisions (American Evaluation Association, 2003).

In recent years, as a result of NCLB and PART, questions have been raised about (1) the sole focus on outcome assessment (with little to no focus on assessing processes), (2) the reliance on external evaluation experts to conduct evaluations, and (3) the mandate to use only rigorous methods, particularly experimental designs, that can produce scientifically based evidence as to the efficacy of program interventions. As a result of concerns raised about the narrow view of what is important when evaluating a program, there have been a number of developments within the field, which provide a look at the other side of evaluative efforts.

Approaches to Evaluation. Generally speaking, program evaluation is a systematic process for gathering and analyzing data to assess an intervention effort implemented to bring about a change in a target group, organization, or society as a whole. The change effort has been implemented to resolve issues or bring about positive changes in the targeted entity. Program evaluation is a planned and purposeful activity that addresses questions about worth and value of that which is being evaluated. Knowledge is gained using a variety of methodological and analytical techniques. Furthermore, this knowledge and understanding is to enhance decision making, particularly as it relates to making improvements in a program, process, product, system, organization, personnel, or policies. Also, the evaluative knowledge can be used to make decisions about the allocation of resources (e.g., whether to continue or expand a program). Unlike traditional academic research, evaluation is grounded in the realities of organizational operations, and evaluation questions arise out of everyday practice.

There are two basic types of evaluation, although the labels attached to each type may differ. Formative (process) evaluation is conducted to assess the implementation of a program (particularly processes) to gain knowledge that can be used to make improvements in those processes. Summative (outcome) evaluation focuses on the results of a programmatic effort and on whether the goals of a program have been achieved. Summative evaluation, by its nature, makes a judgment about a program's worth, merit, or value. In addition to these two basic types of evaluation, an evaluator can play a role in the developmental phase of a program's design to gather information and provide informal feedback to an organization before a program is implemented.

The evaluation field offers a wide range of methodological techniques, including (1) various methods to gather and analyze data (e.g., quantitative, qualitative, and mixed methods), (2) use of different types of research designs (e.g., naturalistic and experimental), and (3) attention to various foci (e.g., processes, outcomes, impacts, costs, and cost-benefits, to name a few).

There are a number of evaluation approaches that have been developed over the years that have implications for the process improvement model articulated in this book and that cross over all these methodological techniques. These include approaches that emphasize the importance of organizational learning, utilization-focused evaluation, and stakeholder involvement in the evaluation effort (e.g., participatory, collaborative or empowerment evaluation and use of appreciative inquiry methods).

As a concept relevant to organizations, organizational learning was first discussed in the work of Argyris and Schön (1978, 1996), where they distinguished between single-loop and double-loop learning. Single-loop learning is when individuals, groups, or organizations modify their actions based on the differences between expected and obtained outcomes. Double-loop learning is when the values, assumptions, and policies that led to the actions in the first place are questioned and modified. This represents a learning process that results in changes in organizational values and promotes a fundamental transformation within an organization. As argued by Argyris and Schön, when organizations engage in double-loop learning they are more productive and successful than when they engage in single-loop learning.

More recently, organizational learning has been popularized by Peter Senge (1990). His focus is on group problem solving and how systems thinking can convert companies into learning organizations. In his book, The Fifth Discipline, Senge describes five characteristics (disciplines) of learning organizations: (1) systems thinking (which integrates the other four), (2) personal mastery, (3) mental models, (4) shared vision, and (5) team learning. The five disciplines provide a conceptual and methodological basis for developing three core learning capabilities: (1) fostering aspiration, (2) developing reflective conversation, and (3) understanding. Engaging individuals within an organization is central to the change process, as reflected in Senge's notions of personal mastery (having individuals committed to lifelong learning), mental models (the capacity of individuals to scrutinize and share their own thinking and assumptions), and shared vision (the fusion of individual and organizational vision). In essence, successful learning organizations do not only rely on the commitment to learning on the part of the organization, but also recognize the centrality of individuals within an organization and the impact they have on the change process (Owen, 2005).

In an average lifetime, a person walks about sixty-five thousand miles. That's two and a half times around the world. I wonder where your steps will take you. I wonder how you'll use the rest of the miles you've been given.

— Fred Rogers

A learning organization is able to create, acquire, and transfer knowledge, as well as modify its own behavior, to reflect new knowledge and insights. Successful learning organizations are able to use data on a regular basis, assess its implications, and make the necessary changes based on their analysis. The building blocks of a learning organization are systematic problem solving, learning from experience, and the transfer of knowledge through a variety of mechanisms (Garvin, 1993). Building on what works and changing continuously in an effort to be more effective is what organizational learning is all about. To engage in an organizational learning process to become more effective, organizations must have very clear goals, priorities, and measures. Furthermore, there needs to be an alignment of goals and measures for people, for budgeting and financial management, and for assessing organizational results. Successful learning organizations also have

1. leadership that communicates and motivates;
2. usable and accessible information management systems that give managers and teams data about what is working;
3. more transparency with respect to what is going on—not only with managers, but also with the people being served, and others;
4. trust in team efforts, allowing creativity in problem solving, encouraging risk-taking, and recognizing that failure is an opportunity for learning; and
5. opportunities to recognize and celebrate successes, particularly of teams and the whole organization (Chapel & Horsch, 1998, p. 9).

Regardless of the differences among theoretical discussions about organizational learning, they all imply that change is a continuous process and involves elements of reflection and analysis. Figure 2.3 provides a model for a five-step organizational learning process, as articulated by Weiss and Morrill (1998). This model, of course, is similar to the PDCA cycle in the quality control and lean literature and to the continuous change cycle in Figure 2.2. It is obvious that while there might be some terminology differences across disciplines, the fundamental notion is the same: change must be a continuous process of reflection, analysis, and improvement based on knowledge acquired as a result.

Utilization-focused evaluation refers to the way in which findings from an evaluation are used and how programs or organizations change as a result of these findings. As maintained by Patton (1997), evaluations should be judged by their utility and actual use. Evaluators should design their evaluations with careful consideration as to how the design and conduct of the evaluation will affect use.

The "use" of an evaluation is defined in terms of how real people in the real world apply evaluation findings and experience the evaluation process. Therefore, the focus in utilization-focused evaluation is on intended use by intended users. Since no evaluation can be value-free, utilization-focused evaluation answers the question of whose values will frame the evaluation by working with clearly identified, primary intended users who have responsibility to apply evaluation findings and implement recommendations (Patton, 2002, p.1).

The issue of utilization raises questions about the role of the evaluator and intended users of evaluation findings. There is a psychology of use that provides insight into how evaluators should approach the entire process from beginning to end (Patton, 2005, p. 429).

1. Intended users are more likely to use evaluation if they understand and feel ownership of the evaluation process and findings.

FIGURE 2.3. Organizational Learning Process

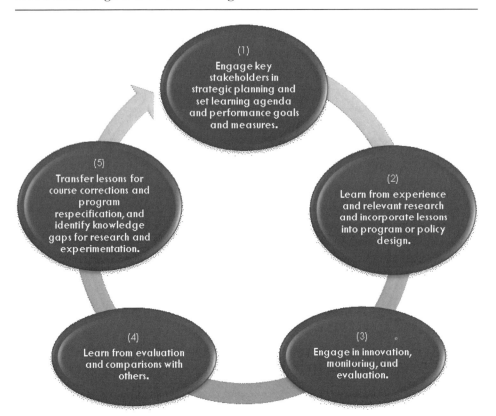

2. Intended users will feel ownership if they have been actively involved in the design process.
3. Evaluators are responsible for paving the way for users to understand evaluation, laying the groundwork for its use, and reinforcing its utility throughout the process.

Another key concept related to utilization concerns the identification of potential users of evaluation. Those within the field have adopted the term stakeholders, since there are any number of potential users, but the commonality is that stakeholders are people who have a stake or vested interest in the evaluation findings. Persons who are stakeholders include (1) people who have decision-making authority over a program (e.g., funders, policy makers, advisory groups), (2) people who have direct responsibility over the program's operation (e.g., program developers, administrators, managers, and direct service staff), (3) people who are beneficiaries of a program (e.g., participants, their families, and communities), and (4) people disadvantaged by a program, such as those who have lost funding opportunities because of the program (Green, 2005). In an even broader context, stakeholders might include journalists, taxpayers, and the general public, particularly with respect to publically funded programs.

The challenge for evaluators is to be attuned to what may be competing interests of the various stakeholders. Evaluation efforts cannot address the questions of all the various stakeholders; therefore, it is necessary for evaluators to focus in on primary users of evaluation. Given this, each evaluation situation is unique, with a mixture of "people, politics, history, context, resources, constraints, values, needs, interests, and chance" (Patton, 2005, p. 430).

Discussion about the importance of stakeholder involvement in the evaluation process has led to other caveats as the evaluation relates to the purpose of and approach to evaluation. While we have already pointed out one rationale for involving stakeholders (to foster the utilization of evaluation findings), a secondary rationale is to advance the values of equity, empowerment, and social change within the context of the evaluation. This rationale for stakeholder involvement grew out of the influence of participatory action research in the 1970s; this research was practiced in the developing countries around the world where there was a concern about issues of justice, equity, and empowerment for those stakeholders without power who were the beneficiaries of programmatic efforts.

An overarching term used to describe an approach to evaluation involving stakeholders is participatory evaluation. In general, participatory evaluation involves program staff and clients (generically referred to as "participants") in the planning and implementation of evaluations. There are several characteristics of this approach to evaluation (King, 2005). First, the direct and active participation of participants is over time, and is more than simply providing a source of data and information about how a program operates and what its results are. Second, the role of the evaluator is that of partner, facilitator, or coach who ensures the active participation of the participants in the evaluation process. Once the capacity of the participants has been developed, the responsibilities for evaluation may ultimately devolve to them. Throughout this facilitation process, it is critical to have activities involving the participants so they establish their ownership of the evaluation, which is then likely to increase their utilization of the evaluation findings.

Often the terms collaborative evaluation and participatory evaluation are used interchangeably. To distinguish them, however, collaborative evaluation is one type of participatory evaluation. Collaborating is one way to work together, and implies working together as co-equals (King, 2005).

Another form of participatory evaluation, but a form with a different purpose, is empowerment evaluation. Its use implies the use of evaluation concepts, methods, and findings as a means to foster improvement and self-determination on the part of the organizations, programs, or people being evaluated (Fetterman, 1994). The underlying rationale for empowerment evaluation is to help people help themselves. Therefore, this type of evaluation focuses on self-evaluation and reflection as a means to ensure that clients, consumers, and staff (program participants) are able to improve programs on their own. Thus, the role of an evaluator in empowerment evaluation, as with other participatory evaluation models, is to act as a facilitator and coach, with the ultimate goal to build the capacity of program participants. The participants internalize and institutionalize self-evaluation processes that are ongoing and that lead to program improvements over time.

In addition to these various models of evaluation that have stakeholder involvement, another important development in the field of evaluation that is relevant to this discussion, is appreciative inquiry. In this approach to inquiry, the focus is on understanding what is best about a program, organization, or system, and what will lead to the development of a better future. Underlying assumptions of appreciative inquiry include these (Preskill, 2005, p.18): (1) What we focus on becomes our reality. (2) There are multiple realities and values that we should acknowledge. (3) The act of asking questions influences perceptions and behavior. (4) People will have more enthusiasm and

> *The big secret in life is there is no secret. Whatever your goal, you can get there if you're willing to work.*
>
> — Oprah Winfrey

motivation for change if they see possibilities and opportunities for the future. The five principles upon which appreciative inquiry is based are these:

1. Knowledge about an organization and the destiny of that organization are interwoven.
2. Inquiry and change are not separate but are simultaneous. Inquiry is intervention.
3. The most important resources we have for generating constructive organizational change or improvement are our collective imagination and our discourse about the future.
4. Human organizations are unfinished books. An organization's story is continually being written by the people within the organization, as well as by those outside who interact with it.
5. Momentum for change requires large amounts of both positive affect and social bonding—things such as hope, inspiration, and sheer joy in creating with one another. (Preskill, 2005, pp. 18–19)

Limitations of Current Evaluation Approaches. While current evaluation approaches and methodologies focus on both process and outcomes of programmatic efforts, they often fall short with respect to identifying and giving guidance on process improvement strategies. Evaluations of human service programs seldom do a deep analysis of processes and their implementation. Process evaluation questions generally address whether or not program activities were implemented as intended, if the quantity of activity or services was delivered, and perhaps whether the quality of activity or service was acceptable (e.g., were the clients satisfied with the services received or did the services delivered meet best-practice standards). Thus, while a program may deliver both the quantity and quality of activity expected, there still can be significant inefficiencies in a program's operation and a better use of scarce resources to achieve the intended outcomes and long-term impact.

Given the above, the process improvement model for service organizations using lean concepts and methods, as adapted from business and industry, offers another conceptualization and set of methodological tools that can enhance many of the established evaluation approaches. This is particularly relevant when evaluations include a formative component where the results of the evaluation play a role in helping organizations improve their programs. Lean concepts and methods can be an instrumental part of an evaluator's toolkit because they provide a way to rethink our approach to evaluation and offer practical tools to improve the implementation of processes that are robust and that ensure the achievement of intended outcomes.

QUALITY IMPROVEMENT MODELS

The quality movement in the United States emerged in the 1980s; it has been one of the most pervasive and significant developments in business practice (Stecher & Kirby, 2004). Whether known as Total Quality Management (TQM), Continuous Quality Improvement (CQI), Continuous Improvement (CI), or Quality Management (QM), the essence of quality management philosophy and set of practices include these:

1. Continuous improvement
2. Meeting customers' requirements
3. Reducing rework

4. Long-range thinking
5. Increased employee involvement and teamwork
6. Process redesign
7. Competitive benchmarking
8. Team-based problem solving
9. Constant measurement of results
10. Closer relationships with partners or suppliers (Powell, 1995; Ross, 1993)

The key thinkers and their frameworks for continuous quality improvement, which are particularly relevant to our application of lean thinking to service organizations, include the work of W. Edwards Deming, the Malcolm Baldrige National Quality Award Program, Six Sigma, and Theory of Constraints.

W. Edwards Deming

The U.S. focus on quality and continuous improvement grew out of a response of U.S. manufacturing companies that were facing strong global competition from Japan, which had revolutionized its national industrial business practices after World War II. The revolution in Japan was greatly influenced by the work of W. Edwards Deming, an American statistician, professor, author, lecturer, and consultant, however. Deming trained hundreds of Japanese engineers, managers, and scholars in statistical process control and concepts of quality. As a result of Deming's work in Japan, manufacturers there increased their quality and productivity to unheard-of levels, resulting in a new international demand for Japanese products. This was the basis of the Toyota Production System and Lean Manufacturing discussed in Chapter 1.

Deming's philosophy focuses on the adoption of key principles of management, which are drawn from his System of Profound Knowledge. That philosophy includes appreciation of a system, knowledge of variation, theory of knowledge, and knowledge of psychology (Deming, 1986, 1993). Of critical importance in Deming's work is planning for the future and having a long-term commitment to new learning. Over time, this enables organizations to increase their quality and reduce their costs through reducing waste, rework, staff attrition, and litigation, while simultaneously increasing customer loyalty. For Deming, it is essential to practice continual improvement and think of organizations in terms of systems, not as separate bits and pieces.

Deming's work also influenced business and industry in the United States, although some may argue that, as a nation, we still suffer from our focus on short-term or immediate results, reliance on technology to solve our problems, and use of quick fixes rather than using deep analysis and problem solving. Regardless, Deming's work did have an influence on the growth of total quality management in the United States in the 1980s.

The Malcolm Baldrige National Quality Award Program

The total quality movement gained momentum in 1987 when the U.S. Congress established the Malcolm Baldrige National Quality Award Program (MBNQA), which created criteria for performance excellence and provided organizations with a framework for designing, implementing, and assessing a process for managing all business operations to be able to meet those criteria (Kirby, 2004). The U.S. Commerce Department's National Institute of Standards and Technology (NSIT) manages the MBNQA program, with assistance from the American Society for Quality. This program is a joint government and private sector effort that has expanded to other sectors in recent years.

> *A funny thing about life: if you refuse to accept anything but the best, you very often get it.*
>
> — W. Somerset Maugham

29

The award recognizes excellence in manufacturing, service, small business, education, and health care.

The excellence criteria emanate from a set of core values and concepts, including visionary leadership, customer-driven excellence, organizational and personal learning, valuing employees and partners, agility, focus on the future, managing for innovation, management by fact, social responsibility, focus on results and creating value, and a systems perspective. The seven criteria that form the basis for organizational self-assessments and making awards are (1) leadership, (2) strategic planning, (3) customer and market focus, (4) information and analysis, (5) human resource focus, (6) process management, and (7) business results. It is of critical importance in this award program that organizations establish a results-oriented framework, which creates a basis for action and feedback (National Institute for Standards and Technology [NIST], 2001).

Research that assessed the relationships between the MBNQA categories has confirmed the appropriateness of the criteria as they relate to organizational outcomes (Pannirselvam & Ferguson, 2001; Powell, 1995; Winn & Cameron, 1998). Findings suggest that leadership significantly affects the other criteria. In addition, this research points to the importance of having a strong customer focus, since implementing other quality management procedures will not be successful, in and of themselves, without a focus on customers. According to Powell (1995, pp. 21–22), these factors are required for total quality management programs to be a success:

1. a culture receptive to change;
2. a motivation to improve,
3. people capable of understanding and implementing quality management's set of practices;
4. corporate perseverance;
5. leadership capacity to commit; and
6. an exogenous chance factor that may motivate change and learning.

Furthermore, Powell identifies executive commitment, open organization, and employee empowerment as essential ingredients of a culture of quality improvement and as critical to quality management success. Other tools used in quality management, such as benchmarking, training, flexible manufacturing, process improvement, and improved measurement play smaller roles, according to Powell.

With respect to whether investing in quality principles and performance excellence pays off, the empirical evidence is mixed. While there may be evidence of better employee relations, improved product quality, lower costs, and higher customer satisfaction, for some organizations there are negligible gains in profitability (U.S. General Accounting Office, 1991). Furthermore, there have been unanticipated side effects of successful quality programs, as reported by Sterman, Repenning, and Kofman (1997). Based on their study of Analog Devices, Inc., a leading manufacturer of integrated circuits, a quality management program can raise productivity and lower costs in the long term. In the short term, however, a business also can experience excess capacity, financial stress, and pressure for layoffs. Thus, it is important to recognize the complex relationship between successful improvements and financial performance.

Six Sigma

Six Sigma is a process improvement methodology that focuses on the reduction of variation in a process. In statistics, a "sigma" refers to the standard deviation from the mean of a population. The standard deviation is a measure of the likelihood of a data point deviating from the mean of the whole data set. The sixth sigma refers to the likelihood that only 3.4 out of every 1 million data points will appear outside the sixth standard deviation (Bizmanualz, 2008). In terms of business metrics, this translates into fewer than four errors per 1 million transactions, enabling businesses to reduce the variation in the outcomes of their processes. In essence, variation in a process is all about waste (i.e., waste that results from error). Therefore, organizations using Six Sigma aim to understand the process elements so they have more control over their processes and have greater predictability over the results. The assumption is that the outcome of a process will be improved by reducing the variation in multiple elements (Nave, 2002). Six Sigma is problem focused and uses a scientific approach called DMAIC to analyze a specific problem. Again, DMAIC is another variation on the continuous improvement cycle as depicted in Figure 2.2 and discussed earlier in this chapter.

Variance reduction is important in high-technology manufacturing (to improve manufacturing yields), high-transaction businesses (check or postal processing), or environments where errors are expensive (surgical operation, space exploration, aircraft takeoff and landings). Six Sigma uses a set of statistical tools; through a rigid and structured methodology, its application of those tools enhances identification of elements that influence a process, which enables management to establish a process where they have more control over the process and are better able to predict the expected outcomes (Bizmanualz, 2008).

This type of analysis to reduce variation has secondary benefits in that it introduces the evaluation of VA status of process elements and identifies elements that may constrain the flow of products or services through a system. Once identified, changes based on this knowledge can be made in processes, with an overall improvement in the quality of the product or service (Nave, 2002).

Theory of Constraints

The theory of constraints (TOC) is an overall management philosophy based on the application of scientific principles and logic reasoning to help organizations continually achieve their goals. It focuses on system improvement, where systems are defined in terms of a set of interrelated processes (Dettmer, 1997; Goldratt, 1994). According to TOC, every organization, at any given time, has at least one constraint that limits the system's performance relative to its goal. The critical question asked is, How do constraints affect performance? Generally, a constraint is viewed as a bottleneck, a delay, or a barrier to reaching the full potential of a system. Therefore, to manage the performance of the system, the constraint must be identified and managed correctly. Once the goal of the organization is articulated, the TOC specifies five focusing steps that are required to implement an effective process of ongoing improvement. These steps include the following:

1. Identify the constraint or whatever prevents the organization from obtaining its goal.
2. Decide how to exploit or maximize the efficiency of the system's constraint. In other words, make sure the constraint is doing things the constraint uniquely does, and that it is improved or otherwise supported to achieve its utmost capacity without major expensive upgrades or changes.
3. Subordinate other processes to the constraint. In other words, align all other processes to the constraining process that is working at maximum capacity. To accomplish this alignment, the other processes must be paced to the speed or capacity of the constraint.

4. Elevate the constraint if the output of the overall system is not satisfactory. This may require further improvement by making major changes to the constraint that involve capital improvement, reorganization, or other major expenditures of time and money.

5. Repeat the cycle of improvement. In other words, if the first constraint is fixed and the constraint has moved, the focus turns to the new constraint and the cycle is repeated.

The TOC provides a methodology aimed at improving the flow time of a product or service through a system. Specifically, when there is a reduction of waste in the constraint, it improves throughput; when the constraint is improved, it reduces variation. Overall, this methodology improves quality and the performance of a system to reach its goal.

In addition, the TOC identifies a set of tools (thinking processes) to help managers apply this method of process improvement and gain buy-in, thereby reducing any resistance to change. The thinking process includes these points: (1) gain agreement on the problem, (2) gain agreement on the direction for a solution, (3) gain agreement that the solution solves the problem, (4) agree to overcome any potential negative ramifications, and (5) agree to overcome any obstacles to implementation.

Quality Models and Improving Organizational Performance

The quality improvement models discussed above do not constitute all the various ways in which quality within a business or organization can be managed. Quality management, as a concept, evolved during the early stages of the industrial revolution as attempts to improve efficiency, quality control for production, and standardization of processes were of paramount importance.

Over the years, the literature on quality has grown extensively. The emerging themes relate to the ability of an organization to develop a quality culture and the role of leadership in promoting and achieving high quality. Moreover, through systems thinking, a more holistic approach to quality is proposed—one that considers people, process, and product or service together, rather than as independent factors in quality management. In addition, a new paradigm is emerging that takes quality as a given within organizations. For businesses and organizations to excel and be successful, they must move to a stage of innovation management, where the focus is on how the organization uses its creative efforts to introduce new ideas, processes, or products (Trott, 2008).

Our discussion of the quality models and our discussion of approaches to program evaluation offers a glimpse of the various conceptual and methodological perspectives related to improving organizational performance. Together, they provide the context within which we can articulate the proposed model of improving performance through a lean transformation, as adapted to service organizations. This context has been provided by describing the cross-fertilization of concepts and methods that has occurred within business and industry, the applied social sciences, and governmental initiatives.

The methods proposed herein are not new with respect to ways to improve organizational performance. However, service organizations have yet to grasp the power of analyzing their processes through a lean lens to identify areas of waste. This analysis could lead to an optimization of their organizational performance. Our book fills a gap in the literature and provides a way to translate lean concepts and methods in a way that makes sense to service organizations.

Opportunities multiply as they are seized.

— Sun Tzu

PART I

From Knowledge to Practice

Part I provides an introduction to lean thinking, with a focus on its application to service organizations. Specifically, Chapter 1 provides an overview of improving performance through a lean transformation and identifies the benefits for your service organization. Chapter 2 establishes a context for understanding why it is important to improve organizational performance through an examination of organizational systems, the drivers for performance-based accountability and program evaluation, and quality improvement models. The following exercises provide you with the opportunity to take the knowledge gained in this section and put it into practice.

    Exercise I.1. **Regulatory Requirements and Standards of Best Practice**
    Exercise I.2. **Performance Assessment Within Your Organization**
    Exercise I.3. **Developing a Logic Model**
    Exercise I.4. **Evaluation Models and Continuous Quality Improvement**

In light of our emphasis on learning by doing, the final activity in Part I asks you to reflect on the exercises completed and engage in a discussion about the challenges associated with completion of each exercise and any lessons learned as a result of your efforts that can help you in further attempts to apply this knowledge.

Exercise I.1. Regulatory Requirements and Standards of Best Practice

Identify any regulatory requirements or standards of best practice established by an outside authority (e.g., federal government, professional association, accrediting body) that impact your service organization. Using these as a point of departure, identify several performance requirements specific to your organization and your responsibility for tracking and reporting the performance measures.

Exercise I.2. Performance Assessment Within Your Organization

Does your organization assess processes in a way that is congruent with lean thinking? Discuss the ways in which you think this approach to improving performance can be applied within your organization.

Exercise I.3. Developing a Logic Model

If you have a logic model for a program within your organization, identify two to three performance-related questions that you can extract, which can be used as a basis for developing performance measures and their specific indicators. If you do not have a logic model, use the example in Appendix A to complete this exercise.

Exercise I.4. Evaluation Models and Continuous Quality Improvement

Taking the approaches to evaluation and quality improvement discussed in Chapter 2, identify at least five underlying principles that you think are important in the assessment of organizational processes and the development of a system for continuously improving them (e.g., stakeholder involvement). Provide a discussion as to why you think these principles are important.

Reflections and Lessons Learned

1. Reflect on your effort to complete the exercises you completed in Part I, and discuss the challenges and any lessons learned associated with that effort.

2. Identify what you think might be a better set of exercises or questions to answer that would help you critically think about and better understand the context within which lean philosophy can be applied to service organizations as a means to improve their performance and instill a culture of organizational learning.

Part II Overview

Understanding Basic Concepts of Lean Thinking

To their credit, the evaluation and process improvement models discussed in Chapter 2 emphasize the importance of continuously improving organizational systems and using systematic approaches to understand what an organization is doing, how well it is doing it, and what changes need to be made to improve the organization's systems. However, some of these models are long on abstract thinking and short on real world application. It is important to delineate a conceptual framework and set of methodological tools that incorporate proven approaches to improving performance and offer a set of practical applications that you can readily use within your service organization.

It is with this in mind that we provide an adaptation of lean concepts that can be used in your service organization, taking into consideration the approaches to understanding organizational behavior and quality improvement processes as discussed in Part I of this book.

Chapter 3 introduces the concepts of value streams, wasteful activities, and unacceptable results. The concepts of value streams and wasteful activities lay the foundation for examining processes through a lean lens. These are con-

cepts that are unique to lean thinking and provide a way of looking at both organizational efficiency and effectiveness. Chapter 3 provides definitions and detailed descriptions of these concepts that make them relevant to organizations providing services, rather than to those producing products in a manufacturing environment. This chapter also presents a new concept—unacceptable results (URs)—to articulate problems or issues that arise with the way in which organizations conduct their work. Once your organization identifies the URs that you are experiencing, the next steps for creating a more-efficient and more-effective system can take place.

Chapter 4 takes another look at performance measures, particularly as they relate to the use of those measures in lean transformations. Since there can be many levels of measuring performance and timeframes (e.g., short term versus long term), this chapter provides guidance for designing performance measurement systems that will keep your cycle of continuous improvement on track, without long delays while you wait for the results from changes you made in operational processes.

CHAPTER 3

Value Streams, Wasteful Activities, and Unacceptable Results

CHAPTER 3 AT A GLANCE

VALUE STREAMS
- Value Streams in Service Organizations
- Key Features of Value Streams
- How to Identify a Value Stream
  - Service or Product Families
- Factors Used to Isolate a Value Stream
  - Client Groups
  - Work Volume
  - Client Concerns
  - Business Conditions

VALUE-ADDED VS. WASTEFUL ACTIVITIES
- How to Assess Value
- Non-Value-Added (Wasteful) Activities
  - Waiting
  - Convoluted Pathways
  - Rework
  - Information Deficits
  - Errors or Defects
  - Inefficient Work Stations
  - Extra Processing Steps
  - Stockpiled Materials or Supplies

- Excess Services or Materials
- Process Variation
- Resource Depletion

UNACCEPTABLE RESULTS
- Essential Features of a Lean Organization: FAQWOES
- Indicators of FAQWOES
  - Flow of Work
  - Adequacy of Resources
  - Quality of Services and Products
  - Workload Balance
  - Organization of Work Environment
  - Effectiveness of Process
  - Standardization of Work

HOW TO ASSESS WASTEFUL ACTIVITIES AND UNACCEPTABLE RESULTS
- Measurements
- Units of Analysis
- Data Collection Methods

This chapter presents a number of key concepts, providing you with a framework for understanding the basics of lean thinking and how this perspective can produce significant improvements and enhanced performance within your organization. The concepts of value streams and WAs are an integral part of lean thinking; we present these concepts in this chapter. In addition, we present the concept of URs as these are seen or experienced by staff, clients, or other stakeholders. Because these results are seen as problematic by these individuals, they are indicators of process design and implementation issues. Your organization's management and employees must have a deep understanding of these concepts and become accustomed to pinpointing them in your processes on a routine basis.

VALUE STREAMS

While the concept of a value stream was initially defined and used to describe the manufacture of a product, we can apply this concept to your service environment, where you deliver a service (e.g., health care, education, or counseling services, to name a few) or produce other tangible deliverables for your target or client group(s), such as health-care equipment/devices, educational materials, reports, and so forth.

Value Streams in Service Organizations

In your service organization, a value stream includes all the steps or series of actions from beginning to end that are required to deliver services or produce deliverables for your clients. The end of the value stream is the point at which your clients receive the services or other deliverables.

Using the metaphor of a river or a stream (see Figure 3.1), water flows from its initial source in a mountaintop lake and continues its journey through various land formations until it reaches its end point of an ocean, lake, or sea. Some rivers encounter treacherous landscapes along the way, which are hindrances to a smooth flow of the water. Within your work environment, the stream represents the flow of work that begins upstream with a request for service to meet your client's need. The downstream flow represents the work processes that occur to deliver the service to your client. The end

FIGURE 3.1. Two Rivers: One With Hindrances, One Without

point is reached when your client receives the service. However, just as rivers have lots of twists and turns and other hindrances that impact the even flow of water from beginning to end, your work environment may have its hindrances, as well. These hindrances result in NVA activities in that they do not contribute to the delivery of service to your client. These NVA activities should be eliminated so that the flow of work is smooth and your client receives a high-quality service without delay (Tapping & Dunn, 2006).

As depicted in Figure 3.2, we can take this metaphor and use it to describe a value stream within a health-care facility (e.g., radiology services). We start at the end point to identify the client (i.e., patient in this case) and the purpose for which the radiology department exists for a patient. In this situation, the radiology department conducts an assessment of a specific condition and produces a report that goes to the patient's referring physician. To identify the starting point, we must walk backwards through the value stream to the farthest point that is relevant for the service received by the patient. Referral by the patient's primary physician is the logical starting point. After a referral, the patient will move through the processes of intake, prep, procedure, and reporting that make up the entire value stream. These are the macro steps within the value stream, which are the points of departure when value stream mapping is completed, as will be described in Chapter 5.

Key Features of Value Streams

The following are key features of value streams that are important to keep in mind as we discuss their use in describing, analyzing, and improving work processes within your service organization:

1. The value stream for a service or other deliverable within your organization flows to some specific end. This end consists of outcomes or results for your clients that have a set of requirements associated with them (e.g., the level of improvement or quality expected with respect to your clients' conditions, attitudes, behaviors, etc.).

FIGURE 3.2. **Value Stream for Radiology Assessment Within a Health-care Facility**

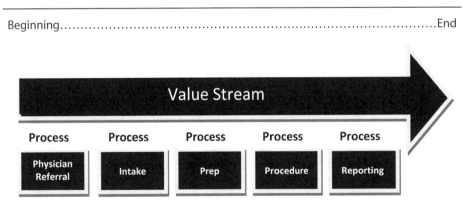

This value stream includes all the major steps that are required from beginning to end for a patient to have a radiology assessment completed, with a report sent to the patient's referring physician.

2. The steps in your identified value stream can be described as either VA or NVA with respect to meeting the needs of your clients or achieving the expected outcomes.

3. There may be internal or external clients associated with your value stream. The internal clients within your value stream are interested in the value that flows to them from upstream process steps, while the external clients are interested in the value that flows to them in the delivery of your service or product.

4. If your value stream fails to meet the set of requirements or expectations of your clients (either internal or external), then it must be improved.

5. The elements flowing through your value stream include individuals or materials that are being transformed as they go through the process steps, along with the flow of information that guides the process steps. (In your service environment, the transmission of information may be the deliverable. When information is the deliverable to your client, it is different from the information that flows through your value stream to guide the movement of the deliverable. It is important to keep these two types of information separate.)

6. Your value stream may be a single process or a linked series of processes. Typically, a value stream connects a single customer through multiple processes.

7. The goal is to create continuous flow in your value stream (i.e., the individuals or materials never stop as they move through the value stream).

8. The flow in your value stream is based on "pull" rather than "push" factors (i.e., movement through the value stream comes from client needs or requests; those needs or requests pull the individuals or materials through your value stream to meet those needs or requests). For those needs or requests to be met, information about your client needs and requests must move back upstream and signal each upstream step as to the immediate needs to meet client requests. Ideally, the information about client needs and requests can be supplied quickly.

How to Identify a Value Stream

The beginning and ending points of a value stream can extend beyond the boundaries of your organization. However, it is best to start within your organization if the control and authority of your organization's leadership or other change agent does not extend to external organizations and processes. Furthermore, for initial improvement projects, we recommend that you begin with a single, less-complicated value stream so your understanding and skills for mapping and analyzing the value stream can be learned before you attempt more-complicated process improvement initiatives.

Service or Product Families. There are different methods used to identify a value stream that will be the focus of subsequent mapping and analysis. To identify a value stream, you must first understand the concept of service or product families. A service or product family is the unit of analysis for mapping value streams. A family consists of a set of services or products that pass through a similar set of processing steps from beginning to end of a value stream, although there may be some differences in the steps. This service or product family is an important concept especially if, within your organization, there are different work units that share the same set of processes. If so, it will be important to take a broader look at how these same processes are completed within different work units to learn about their respective problems and issues. Furthermore, since standardization is one of the key features with any lean transformation, once your process improvements have been made within one work unit, this learning and improvement can be applied to other work units.

To illustrate what we mean by a service or product family, we need to revisit Example 1.1, which described a government economic development agency. Again, the mission of this economic development agency is to empower businesses and communities to invest, succeed, and thrive in an environment that affords a superior quality of life and increases opportunities for economic prosperity for the state. To accomplish this mission, the agency had a number of program areas and offices (i.e., work units) that administered more than eighty different grant programs. These work units provided grants to applicants (businesses or communities) with the purpose of improving their economic stability and prosperity. To determine the service or product family the first step is to create a family matrix, as shown in Figure 3.3., by following these steps:

1. In the far-left-hand column, list the different grant programs; each of these grant programs represents a different client group, based on the type of grant received.
2. Next, identify the number of clients (or volume of service) for each grant program within a specified period (e.g., one year).
3. Across the top of the chart, identify in sequence the different processes that involve all of the grant program areas.
4. Next, put an "x" in the cells to indicate the processes used within the specific grant programs.
5. Finally, group all of the grant programs that have common processes.

This grouping identifies the value stream for the family of grant programs that have many processes in common, although the programs may be administered within different departments. Once the value stream mapping begins, as will be described in Chapter 5, data associated with the work processes carried out in these grant programs are used to add key performance measures to the value stream map (VSM).

Factors Used to Isolate a Value Stream

There may be circumstances within your organization that have a bearing on the factors used to isolate a value stream that will be the focus of your process improvement efforts (Tapping & Dunn, 2006). These factors, as described below, include (1) client groups, (2) work volume, (3) client concerns, and (4) business conditions.

FIGURE 3.3. Service/Product Matrix for Economic Development Agency

	# of applications	Process A Application	Process B Contracting	Process C Monitoring	Process D Compliance and Enforcement	Process E Reporting
Grant Program A	500	X	X	X	X	X
Grant Program B	495	X	X		X	X
Grant Program C	200	X	X	X		
Grant Program D	25	X		X		X
Etc.	25	X	X			X

Client Groups. Depending on the complexity of your organization, you may have a number of different client groups. If so, you may want to identify a value stream based on the service delivered to one particular client group. For example, a hospital may differentiate its clients in terms of inpatient versus outpatient and focus on a value stream associated with the experience of one of these client groups.

Work Volume. It may be that your organization has a sizable percentage of its work flowing through a set of processes. If so, for process improvement efforts to have the greatest impact on your organization's overall functioning, the determining factor for identifying your value stream is work volume. Take a family service organization that offers behavioral health counseling, health and wellness services, and financial counseling. If the organization's volume of work were 80 percent, 12 percent, and 8 percent, respectively, it would make sense for the organization to focus on the value stream for behavioral health counseling services.

Client Concerns. Another method of identifying a value stream within your organization deals with client concerns or some other mandate that a service be improved (e.g., from your funder or perhaps staff internal to your organization). As an example of this method, consider a government agency that provides immigrant services related to applications for tourist visas, student and temporary work visas, resident alien status, and citizenship. Based on the government agency performance, there may be significant time delays, lost applications, and other types of frustrating experiences that applicants for resident alien status have. In this situation, the value stream associated with applications for resident alien status would take precedence over other value streams.

Business Conditions. Finally, if your organization faces a situation where there is a high level of competition for your services or products, then you will want to focus on the value stream where you are facing loss of revenue unless you continually strive to be the best. In this situation, business conditions in your external economic environment help you determine the value stream(s) to focus on when engaging in process improvement efforts. An example of this method might be a health-care lab that provides blood-testing services to local physicians. If the lab is in a community where there is a lot of competition, then it should focus its attention on providing expedient and accurate results from the blood analysis, and make sure that analysis is competitively priced.

> *Your most unhappy customers are your greatest source of learning.*
> — Bill Gates

VALUE-ADDED VS. WASTEFUL ACTIVITIES

As we discussed earlier, the steps in a value stream can be described as either VA or NVA with respect to meeting the needs of clients or achieving clients' expected outcomes. NVA activities are considered wasteful in that they do not contribute to the delivery of a service or product that has been developed to meet your client needs. Furthermore, there may be steps in a process that are considered NVA from your client's perspective. If these steps are required by some other stakeholder (e.g., a funding organization), then these steps become required non-value-added (RNVA).

However, determining the value of a process step is complicated in a service environment because there are other stakeholders involved, those both internal and external to your organization. Having all stakeholders at the table to map out processes and determine the value of a step can result in many heated discussions, raising questions about value from others' perspectives. Therefore, in this application of lean thinking, it is essential to articulate a modified approach to assessing the value of a process step, allowing the perspectives of different stakeholders to be considered.

> *It is not enough we do our best; sometimes we have to do what's required.*
> — Winston Churchill

As an example of this, we worked with a service organization that mapped a value stream for distributing scholarship dollars to eligible childcare practitioners. The goal of the organization was to improve the practitioners' knowledge and skill through their enrollment and completion of college coursework. In the current state of this value stream, one of the processes involved the movement of a scholarship application through several different staff persons who were responsible for reviewing and signing off on the approval or denial of the request. Each of the staff persons responsible for these reviews and sign-offs perceived his or her work as VA. When challenged with respect to whether these extra steps in the process added value to the client (who was a scholarship applicant, in this case), the staff person would launch a lengthy debate. In this situation, the requirements for the scholarship organization to be held accountable for its use and distribution of scholarship funds meant that a step in the process had to involve another level of review and final approval (making at least one of the sign-offs a required NVA step). In the end, after receiving further clarification from the funding organization with respect to the requirements for oversight of the approval process, the group determined that they had many more sign-offs than were required. As a result, the group reached a consensus that many of the subtasks in this process were NVA, and that the one sign-off was a required but NVA activity.

How to Assess Value

Undoubtedly, reaching a consensus about the value of a process step may not be straightforward. However, actual discussion among various stakeholders is, in and of itself, a valuable exercise and may be more valuable than the actual decision about the label attached to a process step. Heated discussions and exchanges make different viewpoints on program processes apparent. Furthermore, these discussions among the various stakeholders reveal any differences in the criteria used to assess value. In addition, stakeholders who are not involved in the day-to-day operations of a program come to learn about daily operations, with all its concomitant challenges. With good facilitation of the group process, various stakeholders can reach a consensus. Absent good facilitation, we recommend that you label the process step as VA, taking it off the table as a process step that is the focus of improvement efforts, unless at a later time it is reassessed and determined to be NVA.

Non-Value-Added (Wasteful) Activities

Within the lean conceptual framework, NVA activities are a form of waste. For the manufacturing sector, seven deadly wastes have been identified and defined: (1) defects in products, (2) overproduction of goods not needed, (3) inventories of goods awaiting further processing or consumption, (4) overprocessing embedded in the work effort, (5) unnecessary movement of people, (6) unnecessary transport of goods, and (7) waiting by employees for process equipment to finish its work or on an upstream activity (Ohno, 1988). However, some of these forms of waste and the ways they are defined do not directly apply to an understanding of business or service delivery processes that are found in places other than the factory floor.

Given this, we have adapted these forms of waste and added to the list to identify situations where waste is found in process activities prevalent within a service environment. These WAs include (1) waiting, (2) convoluted pathways, (3) rework, (4) information deficits, (5) errors or defects, (6) inefficient work stations, (7) extra processing steps, (7) stockpiled materials or supplies, (8) excess services or materials, (9) process variation, and (10) resource depletion. These WAs may not include all of the types of waste; a number of them are inherently linked, in that one form of waste leads to another (e.g., extra processing steps will result in clients waiting for a service). Regardless, this list provides a starting point for you to identify waste within your service environment, where critical process inefficiencies (1) may not be visible to either your staff or clients, (2) may not be routinely tracked through your organization's performance measures, or (3) may be more tolerated within your service organization than they are in a manufacturing environment that has stringent regulatory standards to meet. These WAs are described below; Table 3.1 offers a number of examples that can be found in service organizations.

It is important to keep in mind that the levels of severity within each of these types of waste can range widely and the consequences of the WAs may either be minor or have extreme consequences for your organization. Regardless, once you identify the forms of waste in your organization, you need to take steps that involve an analysis of the root cause of the waste (as will be discussed in Chapter 6) and the implementation of process improvements to eliminate the wasteful activity (presented in Chapter 7).

Waiting. This WA occurs when individuals (staff or clients) cannot proceed with their work or the next step of a process because they are waiting for other parts of the process to be completed. An example is when a client cannot receive a service with minimum wait time because the process design results in unavoidable delays.

Convoluted Pathways. This WA occurs when there are complicated pathways with many twists and turn that people or material must travel through a value stream. An example is when an application for a service moves back and forth through a number of offices for various signatures or additional information before it can be approved; this delays the client receiving the service.

Rework. This WA occurs when a process is set up in a way that requires the same activity to be done more than once. An example is when there are multiple places a piece of information is stored that requires a person to do the same data entry or filing activity more than once.

Information Deficits. This WA occurs when information is missing that is required for a work process to move forward. An example is when applications come into an agency without all the required information completed, resulting in staff spending time gathering the missing information.

Errors or Defects. This WA occurs when services are delivered or materials produced and rejected because of errors, mistakes, or poor quality. An example is when a complex data management system has been developed to help maintain an organization's ability to track delivery of services or accomplishment of outcomes, and the system lacks quality control on the data entry process, rendering the data unusable due to the amount of error found in it.

Inefficient Work Stations. This WA occurs when a workstation is set up in a way that requires more movement of an individual to complete a set of tasks. An example is when an individual must frequently get up from his or her workstation and go across a room or down the hall to retrieve material.

Extra Processing Steps. This WA occurs when there are numerous steps in a process that do not contribute to the delivery of a service or creation of materials. An example is when a request for service must go through many levels within the organization for review and approval before a decision can be made.

TABLE 3.1. Examples of Wasteful Activities Found in Service Organizations

Wasteful Activity	Examples in Service Organizations
Waiting	• A patient waits in the lobby of a hospital to be escorted to an exam room.
	• A supervisor waits for her assistant to compile some data that are required for her to complete a report.
Convoluted pathways	• A contract requires signatures in a sequential order of multiple authorities located throughout an organization's set of buildings. As a result, the path a contract follows is not sequential, but up and down and back and forth throughout the office grounds.
	• An agency receives an application for service that is incomplete, and it must be sent back to the applicant to complete and resubmit.
Rework	• The data or information about the mailing address of a customer must be entered into an order database and then reentered into a UPS database that is used by the organization for ground shipment because the databases are not linked and certain fields will not autopopulate.
	• An agency uses paper files to hand-count the number and type of services it delivers for a quarterly report because its electronic database to record this information has an error in the way it counts.
Information deficits	• An emergency room takes in patients that do not have a list of their medications, requiring the hospital staff to contact others to determine this information.
	• An application form does not have complete and clear instructions, so many applications being received are filled out incorrectly.
Errors or defects	• An agency prints off 10,000 copies of a promotional flyer, only to find out later that there was an error on a critical date on the brochure, requiring it to be reprinted.
	• A hospital pharmacy fills a prescription for a patient but gives the wrong dosage, resulting in the patient's death.
Inefficient work stations	• An agency has a central filing room where all staff must go to retrieve a client's file to work on.
	• A room set up to assemble a set of educational materials (e.g., books, CDs, flyers, and a tip sheet) into a box for mailing does not have materials in the correct order for assembly.
Extra processing steps	• A staff person completing an assignment for his or her supervisor procrastinates completing it, then has to go back to the supervisor to clarify what needs to be done.
	• An auditor for an organization that has a government contract reviews all of that organization's subcontractor files to ensure accuracy in documentation, rather than reviewing a representative sample of these contract files.
Stockpiled materials and supplies	• An organization buys bulk supplies and other material at the end of a fiscal year because the money has to be spent, not out of need.
	• Twenty-five copies of a final report are produced, although only ten are needed for distribution, requiring the remaining copies to be stored.
Excess services or materials	• An organization has a standard package of services it offers to its clients who are looking for work, even though at least half of the clients do not need the entire package.
	• A hospital has the practice of ordering specialty consults for a patient who does not need that consult.
Process variation	• An organization's database has many open-ended fields that are completed by the data entry person, where the same items may be entered in different ways (e.g., person's name: some write first name, last name, others write last name, first name).
	• Four staff members within an agency review applications to determine eligibility for a service; one staffer has a set of criteria less stringent than the other staff's criteria.
Resource depletion	• An organization's staff spend a considerable amount of time dealing with clients who are not eligible for a service (e.g., going through a denial process and handling grievances), thereby limiting the time they have to address the needs of eligible clients.
	• A county government has only enough funds to cover the salary of 85 percent of its workforce, requiring the staff to go on unpaid leave for one day a week, reducing the amount of time staff have to complete the work at hand.

Stockpiled Materials or Supplies. This WA is a result of producing more materials than needed and maintaining an inventory of them. An example is when an organization produces 20,000 copies of a brochure or flyer because it is less expensive to print in quantity. Because not all are needed, it becomes dated and cannot be used, resulting in extra handling and storage space being used, cash being inappropriately consumed for unneeded material, and labor being used to move the material around or dispose of it.

Excess Services or Materials. This WA occurs when an organization delivers more services and produces more materials than are needed by clients. An example is when a program offers a standard package of services to clients, regardless of the clients' needs.

Process Variation. This WA occurs when there are no standardized or defined workflow processes in place. An example is when staff members are performing the same task differently, or when there are exceptions to a workflow process that create the need to work around the process and there are no guidelines as to how this workaround is done.

Resource Depletion. This WA occurs when an organization allocates critical resources (e.g., personnel, time, and money) to completing work activities that add no value to the service delivered or materials produced, thereby diverting these resources away from what is needed for VA work activities.

UNACCEPTABLE RESULTS

Essential Features of a Lean Organization: FAQWOES

When waste is embedded within your service organization's processes, it has consequences experienced by your employees, clients, or other stakeholders; we conceptualize these consequences as URs. URs are the consequences of the way work processes are designed and implemented that are viewed as problems or issues. Features considered as essential in a lean organization provide the framework for identifying URs associated with those features. These features include Flow of work, Adequacy of resources, Quality of services or products, Workload balance, Organization of work environment, Effectiveness of process, and Standardization of work (FAQWOES).

Indicators of FAQWOES

Identifying URs within your organization requires the specification of indicators that provide evidence of problems or issues within your work environment and its processes, particularly the set of process steps representing a value stream. These indicators are the red flags often identified by individuals who are frustrated or dissatisfied with a process as they experience it, whether that individual is a staff person performing the process, a client at the receiving end of a process, or another stakeholder who has knowledge of or connection to a process. Many of these red flags are interrelated and may cross over one or more of the factors considered essential in a lean environment. When URs are prevalent in one or more of your organization's work processes, they impact your organization's performance in accomplishing goals and objectives. As a result, those URs become prime targets for further analysis to determine (1) why they are occurring and (2) how the problem or issue can be resolved and the process improved. Furthermore, these URs, as with the types of WAs found in processes, do not necessarily represent an exhaustive list; many of these URs are inextricably linked with one another.

Flow of Work. As discussed previously, in a lean organization the flow of work is even and steady from the beginning of a process through the end, without extra processing steps, numerous starts and stops, periods of extended wait times, or required twists and turns in the pathways that people or material must travel in order to complete the process. The flow of work in your organization should come from the pull of downstream process steps that ultimately are based on your client needs. Your work processes should begin at a point in time when they are needed (i.e., just in time). Indicators of URs associated with the flow of work include these:

1. **Process churn:** When people involved in a process keep going around and around on an issue and cannot seem to make a decision about next steps.
2. **Red tape:** When a process requires a significant number of sign-offs to get the permission to move forward or has an extensive amount of paperwork associated with it.
3. **Constant interruptions:** When workers experience a significant number of interruptions (e.g., from other staff, phone calls, emails, etc.) as they are completing process tasks, which affects their ability to focus on the work at hand.
4. **Lengthy completion time:** When a process takes longer to complete than expected, based on the number of steps in the process.
5. **Excessive delays:** When there are significant delays in completing process tasks due to unforeseen events (e.g., key staff is incapacitated, work is lost, disasters occur, etc.).

Adequacy of Resources. In a work environment, you must have sufficient time, an adequate number of personnel, and the required material items (e.g., supplies and equipment) to complete work requirements and meet your clients' needs. Therefore, the amount of resources within your organization can vary with respect to being adequate or inadequate. The indicators of URs associated with the adequacy of resources include the following:

1. **Lack of supplies:** When the supplies needed for work are not available, because of either a shortage or complete depletion of the supplies.
2. **Insufficient time:** When the time allotted to complete a work process is not sufficient, given the current requirements and volume of work.
3. **Extensive amount of work:** When staff are overwhelmed with the amount of work piling up in inboxes or on "to do" lists.
4. **Unavailability of equipment:** When the right equipment is not available or not in working order when needed to complete a work process.

Quality of Service or Product. As discussed previously, your organization must design and implement processes with the expectation of providing high-quality services or products to meet your clients' needs. Furthermore, these services or products must be devoid of mistakes, errors, or defects. Indicators of URs associated with the quality of services or products include these:

1. **Quality standards not met:** The service or product does not meet the standards or best practices associated with that service or product.
2. **Inconsistency in quality:** When the quality of a service or product is not consistent over time, and is sometimes good, sometimes poor.
3. **Repeated mistakes or errors:** When the same mistakes are made repeatedly in the delivery of a service or product.

4. **Rework requirements:** When staff must spend a considerable amount of time redoing or correcting the mistakes or errors in their services or products.

5. **Client complaints:** When clients complain about a service or product because it does not meet their needs or because of errors or mistakes.

Workload Balance. The distribution of work across your staff members or across steps in a process must be equal and evenly balanced, or else you will have disruptions in the flow of work, resulting in both bottlenecks and extended wait times between process steps. The indicators of URs associated with workload balance include the following:

1. **Unequal distribution of work:** When some staff members have more work to do than others do, making the distribution of work is unequal.

2. **Bottlenecks:** When the flow of work gets jammed up at certain points in a process, restricting the flow and slowing down the movement of people or material through the process steps.

3. **"Hurry up and wait" syndrome:** When certain parts of a process are completed quickly, but the work cannot proceed to the next step because those responsible are not ready.

4. **Long queues:** When there is a long queue or line of people or material that must wait before moving to the next step of a process, based on "first in, first out" (FIFO) rule.

5. **Waiting:** When either staff or clients must wait for an extended period to proceed to the next step of a process, which may happen for a variety of reasons.

Organization of Work Environment. In a lean organization, your work environment must be organized and clean, with items in their proper places so they can be retrieved or found without delay when needed. Furthermore, there must be no safety hazards as a result of cluttered, unclean work spaces. The indicators of URs associated with the organization of the work environment include these:

1. **Can't find:** When staff have to spend time looking for a missing item or person they need to do their work.

2. **Go get:** When staff do not have the frequently used items or people close at hand and they must spend time to go get that item or person.

3. **Disorganized workspaces:** When a work environment is so disorganized that staff spend time looking for items.

4. **Unclean spaces:** When workspaces are not clean, resulting in disease control issues and can leave a bad impression with clients and other stakeholders.

5. **Safety hazards:** When a work environment is unsafe, which can occur when items are out of place and are potential trip hazards, when there are foods left out or other toxic items that are not properly secured or disposed of, when staff are not properly trained to maintain safety precautions, and so on.

Effectiveness of Process. Within a work environment, your processes must be designed and implemented to accomplish their specific purpose and achieve results as intended. The indicators of URs associated with the effectiveness of work processes include the following:

1. **Work around:** When a process has not been designed and implemented to address all possible variations or issues that may arise. As a result, staff or clients must work around a process step to complete their work.

2. **Dead end:** When a staff person reaches a point in a process at which he or she is unable to proceed or complete the work.
3. **Unintended consequences:** When a process does not do what it is designed to do, and results other than those expected occur.
4. **Lack of knowledge:** When staff do not have a full understanding as to how a process works and they do not have answers to questions about their work.

Standardization of Work. Within your organization, when a work process is standardized there are explicit instructions as to what, how, who, and when the steps of a process are implemented, resulting in consistency of service or product delivery. The indicators of URs associated with the standardization of work include these:

1. **Unclear process definition:** When there are no clearly defined steps of a process, assigned responsibilities, and timelines for completing a work process.
2. **Work silos:** When work in a value stream crosses over a number of functional areas and there is no clear understanding across these areas as to who does what and how their work is connected, i.e., "the right hand doesn't know what the left hand is doing."
3. **Run around:** When people either do not get answers or get conflicting answers to questions they have about the why, who, what, where, and when of work processes.
4. **Chaos:** When a work environment is subject to a lot of confusion and disorder and there is no clear direction to take to complete work responsibilities.
5. **Duplication:** When processes are set up in such a way that certain process steps are repeated unnecessarily either by the same person or by separate people.

HOW TO ASSESS WAs AND URs

In assessing WAs and URs within your organization it is important to determine if they exist as an issue or problem and, if so, their level of severity. The methods used to assess these features for each type of WA and UR can vary, depending on your organization's resources and available time to conduct an assessment. Given these constraints, there are three basic factors to consider when determining the assessment of WAs and URs: (1) how to measure their existence, frequency of occurrence, and level of severity; (2) what unit of analysis to use; and (3) how to collect the data.

Measurements

Whether or not WAs or URs exist in your organization is best indicated by a simple "yes" or "no." Frequency of occurrence is measured by counting the number of times a particular WA or UR occurs within an established period of time. Level of severity can be measured in a variety of ways. One way is to design an instrument that provides a subjective assessment by asking respondents to answer questions about the (1) impact on your organization, from a low impact to high impact, on a scale of 1 to 5 or 1 to 10, or (2) significance as a problem within your organization, from very insignificant to very significant, on a scale of 1 to 5 or 1 to 10.

Units of Analysis

There are different units of analysis that you can use when assessing the WAs and URs within your organization. At the highest level of analysis, a specific value stream can be identified (as it crosses over multiple organizations or functional departments) and the data are gathered with respect to that particular value stream. Another unit of analysis can be specific subprocesses within a value stream; these subprocesses are identified and respondents use the specified subprocesses as a point of departure when providing input as to problems that exist, their frequency, and their level of severity. As a note of caution, using your organization as the unit of analysis will result in data collection problems. If you ask respondents to think generically about all of your organization's work processes, it can lead to data that are quite mixed, making it difficult to specify particular processes and associated problem areas.

Data Collection Methods

There are a variety of methods you can use to gather data relevant to WAs and URs. One method is to develop and administer a survey of staff, clients, or other stakeholders. Another method is for you to use key informant interviews, observation of work processes, or an examination of critical documents and records.

There is no one best way to assess WAs and URs within your organization, and it may be appropriate to identify a variety of measurements and methods of gathering data. Chapter 6 provides a fuller discussion of performance measurement systems and the important role they play in your organization's lean transformation.

CHAPTER 4

Performance Measures

TYPES OF PERFORMANCE MEASURES

I am what I do . . . especially what I do to change what I am.

— Eduardo Galeano

An essential component of your continuous quality improvement efforts is to measure performance, starting with a baseline measurement before process improvements have been implemented and continuing thereafter to determine if the improvements result in expected changes in performance. Tracking performance over time ensures that your organization (1) knows where it stands with respect to key performance measures, (2) identifies performance issues, and (3) provides you with the knowledge about whether further process improvements are required to meet the expected levels of performance. Therefore, it is essential for your organizational learning to have an established performance measurement system in place.

Differentiating between efficiency and effectiveness measures, as well as between quantity and quality measures in service organizations are essential first steps. While all of these measures provide a snapshot of organizational functioning, there are subtle differences with respect to what is being measured, the purpose the measures serve, and the intended audience.

Efficiency Measures
In general, efficiency measures capture the extent to which your organizational processes are streamlined—that is, the extent to which they use the smallest amount of

resources necessary to accomplish a goal or end result (e.g., time, labor, and material resources—all of which have a cost associated with them). Measures of efficiency are particularly important for management within your organization. If you are a for-profit organization, you pursue efficiency in order to increase the bottom line. If you are a nonprofit or publically funded service agency, your funders (e.g., government or foundations) are interested in operational efficiency as it represents the best use of what are often scarce resources. Efficiency of processes ensures that resources are not wasted on activities that do not contribute to the mission or goals of your organization. Furthermore, your organization's clients generally consider efficiency to be important. When processes are not efficient, your clients may be dissatisfied and complain about delivery of services.

In the language of lean, there are a number of factors to consider in determining the overall efficiency of a process. These factors include whether or not each process step adds value and is available, adequate, and flexible (Womack, 2006).

Value. The notion of value is critical in the assessment of each process step. The basic analysis determines if a process step (with its associated indicators of time, labor, and material resources) adds value from the perspective of your clients. To determine this, it is essential to know your clients' needs and to understand how your organization meets those needs.

However, as we discussed earlier, it may be a complicated matter to establish the identity of your clients. They can be internal or external to your organization. If you are a service nonprofit organization, your external clients include those individuals or groups you provide service to, as well as a host of other stakeholders, such as advocacy groups, funders, political and governmental entities, and even the public. Your internal clients are those individuals within your organization that are recipients of the information or materials that are passed to them from earlier steps in the process. Once your set of clients is established, then you assess the steps in a process with respect to whether those steps add value to your clients. If your assessment of value produces mixed results, then a consensus must be reached by those conducting the assessment in order to decide whether or not a step adds value.

Generally, any waiting time for the delivery of a service is considered NVA from the perspective of the primary external client. For example, if a patient in a pre-op holding area must wait for hours before being taken into an operating room, then the patient considers this step of the process to be non-value added.

Available. In the assessment of each step in a process, the extent to which the step is available has to do with whether the step can be completed at the time it is needed. When a step in a process cannot be completed at the time it is needed, there are delays, work piles up in queues, and overall processing time is extended because of the waiting time between one or more steps in a process. Examples of a process step being available when needed are prevalent in administrative and office processes within organizations. If your organization has an intake process that must be completed prior to the time when your client receives a service, and this process cannot be completed because the labor or equipment needed to complete the intake is not immediately available or working properly, then client intake cannot be completed and service delivery is delayed. A good example of this is when your organization has a computer system from which information must be retrieved or in which it must be entered so subsequent steps of a process can be completed. If your computer system is down or going through a process of updating, it is not available for its intended use. Your intake process stops and work piles up until your equipment is up and running again. If the frequency of this occurrence is rare, it does not pose a big problem for your organization. If your computer system is often down or its processing time is slow due to the age of your computers, however, this factor represents a significant amount of wasted time within your organization; this wasted time needs to be addressed to improve your organization's efficiency.

Adequate. Assessing whether or not a process step is adequate has to do with determining the extent to which your organization has the capacity (e.g., the resources such as labor, equipment, or space) to complete a process in a timely manner or when needed. Take an example of a hospital emergency room where an order needs to be completed for a patient to be admitted and moved to a room for an extended stay within the hospital. If a staff member puts that order in but there is no open room (i.e., space) on the floor where the patient must be admitted, then the patient must remain within the emergency room until a room opens up and the patient can be moved. If such a situation rarely occurs, an organization is not likely to take steps to increase its capacity to meet client or customer needs in a timely manner. However, if there are chronic capacity issues, then an organization needs to address them to improve service to clients or customers. Also, issues of capacity can be the opposite—i.e., there might be too much capacity within your organization. This means your organization is using up resources (time and money) to maintain a capacity level that is not needed by your clients. Again, a hospital organization is a good example. If a hospital operates at 50 percent capacity according to historical data (i.e., an average of 50 percent of its rooms are used on a daily basis), yet the hospital still fully staffs all its floors, pays for utilities, and so forth based on an expected 90 percent average capacity, then the hospital is expending a considerable amount of resources for services that are not needed.

Flexible. How quickly a step in a process can be switched from one type of service or product to another is an indicator of whether or not a process step is flexible. For example, if your organization has multiple funding streams for a program of services, a flexible monthly invoicing process is one that uses the same basic process steps and computerized program for gathering the input and creating your invoices, regardless of funding entity.

In addition to knowing how quickly a process step can be switched, it is important to assess the cost for the switch. Ideally, the switching should be at low cost. However, in your service environment many of your process steps may be completed by staff rather than by machinery. Therefore, the key issue with respect to flexibility is whether your staff are cross-trained so labor can be flexed to ensure a balanced workload in situations where you have a cyclical volume of incoming work, staff absenteeism, or open positions that have not been filled. Flexing labor to balance workload will help you prevent bottlenecks in the flow of work through steps of a process. Take, for example, an agency that provides case management services through a complement of staff working with clients that have various behavioral health issues. The individual staff are specialized in a particular behavioral health issue (e.g., substance abuse) and have a set of clients with substance abuse as the presenting problem. If a particular case manager leaves the agency, it is important for the agency to have the ability to seamlessly switch the case manager's cases to others' cases. To ensure this seamless transition, it is important for all case managers to have a standardized process for fulfilling their responsibilities (e.g., intake assessments, protocols for treatment given a presenting diagnosis, documentation of counseling sessions, documentation of closing cases, and so forth), so they can take on additional cases until such time that a new case manager is hired. If the agency is unable to do this switching, clients will not receive their counseling services and their needs will not be met.

Effectiveness Measures

The outcomes of a process (or set of processes) that make up a system that has been designed and implemented to achieve an organizational goal or mission are indicators of the effectiveness of that system. A measurement of outcomes documents whether the results of your effort are the results you intended. In service organizations, particularly in today's environment of shrinking resources, there are calls for results-based accountability. Funding agencies, whether public or private, are interested in the results of the programs they fund in order to (1) justify budget and funding expenditures, (2) ensure accountability, (3) optimize the use of resources, (4) establish performance baselines, and (5) understand program effectiveness (McDaniel, 1996).

There also may be multiple purposes and audiences for outcome information. Those audiences that are internal to your organization, such as administrators or front-line staff, need information about the results they are producing, particularly if they are evaluated based on the outcomes or results they achieve. Therefore, tracking outcomes on a regular basis is essential within your organization; the data gathered throughout the implementation of your program provides administrators with the basis for making judgments about what works and what doesn't work. If your program (i.e., a system of interconnected processes) is not working as intended, then it signals the need for you to determine why it is not working and to make changes in the processes to improve your performance in achieving the outcomes expected.

As an example, a certain educational program has a goal to improve the knowledge of a group of professionals. The expected outcome of that program is to increase the knowledge of the professionals who are participating in the educational experience. The program may have a level of knowledge that is required, per targets established in its program design. One target might be that 80 percent of the professionals participating in the educational program show a 50 percent improvement in their knowledge, as measured by a standardized test administered both before and after the professionals participate in the educational program. If this target is achieved, then the program is considered effective in accomplishing its goal.

To determine this effectiveness, there must be a measurement system in place that provides evidence as to the change in knowledge over time. Establishing a baseline measure of knowledge prior to participation in the program of PD is a necessary first step. A similar measurement of knowledge must occur after the participants have been exposed to the program of PD. This measurement provides the comparison of knowledge levels from before to after participation. Before and after measures of knowledge, at a minimum, are required in the measurement system established to assess the effectiveness of its program of PD.

Indeed, there are measurement designs that are more sophisticated and more complex that can be put in place to provide strong evidence of program effectiveness (e.g., basic researchers would use more-sophisticated experimental designs with control and experimental groups to test theoretical hypotheses about cause-and-effect relationships). However, such complex experimental designs are not necessary for organizations that are tracking their outcomes for purposes of program management and improvement efforts. Your organization can establish a measurement system that continuously tracks outcome measures over time. If you implement a process improvement, then it

The only man who never makes a mistake is the man who never does anything.

— Theodore Roosevelt

is critical that you have a measure of outcomes before and after the implementation of the process improvement. In addition to the timing of outcome measures, other issues associated with the measurement system include the specific measures (or indicators) of the outcomes that will be used and who will collect the data. These aspects of outcome measurement systems are pertinent to all performance measures, and will be discussed later.

With respect to your organization's process improvement efforts, determining the effectiveness of your processes is a critical first step in assessing performance. Once you have determined that a process or interrelated set of processes is logically connected to achieve its intended outcomes for your clients, then the next step is to focus on the efficiency of the process(es). In the case of social service programs, this focus requires that your program with its complement of services has been designed and implemented using evidence-based practices, where the effectiveness of a program of services has been established through basic research efforts. Otherwise, your organization could fall into the trap of making ineffective processes more efficient—ultimately is a waste of time and resources. However, to ensure that improvements are not compromising achievement of program outcomes, it is essential to track outcome measurements over-time, although the timing of measurements may be different from the timing for other performance measurements.

Quality Measures

The quality of a service or other materials or products that an organization delivers to its clients is an assessment of the extent to which these deliverables meet a set of standards that are indicators of quality. Standards of quality can be quite complex, as in the case of accreditation standards used by associations or commissions. For programs to be accredited they must meet these established standards that generally have been developed by professional bodies spending considerable time and effort defining what they mean by "quality" and developing the indicators they will use to assess the level of quality. Accrediting bodies develop measurement tools to gather data from a program's documents and records, or from observing and interviewing internal and external clients and other stakeholders. For example, institutions of higher education are accredited by regional accrediting bodies. One of the standards of a quality institution is that its academic programs have learning outcomes that are associated with a set of academic activities (e.g., coursework or other experiential learning activities). The presence of learning outcomes and associated activities in all academic programs within the institution is one of the indicators of a high-quality institution.

Conversely, some measures of quality may be much less complex and not based on an agreed-upon set of standards that are used as indicators of quality. For example, you may ask your clients about their level of satisfaction with the quality of the services they received (e.g., on a scale of very satisfied to very dissatisfied). An assessment of this nature is not based on any specific standards of quality. Rather, individuals responding to your client satisfaction questions are using their own criteria as to what constitutes quality. If we take an educational program and ask the students to rate the overall quality of a course, the criteria used to make that assessment might include the level of difficulty, the level of competence of the instructor, the extent to which students learned what they expected to learn, and so forth. Furthermore, what the students expect and want out of their course experience is highly subjective—one student may indicate a high level of satisfaction if a course is difficult, rather than easy. Another student may do the opposite—i.e., prefer an easy course and indicate a low level of satisfaction if the course is too difficult.

Regardless of how simple or complex your measures of quality, they are important for both your internal and external clients and stakeholders. As an organization, you should want to be known for the delivery of high-quality services or products. Your clients want to receive the same. However, you may have levels of quality that can be achieved only with an expenditure of resources that is not affordable for your organization or your client. If so, then the level of quality you expect will vary depending on those resources. Regardless of the targeted level of quality expected, if your organization collects data on quality showing that a service or product is not meeting that expected level, these data signal a need for improvements to achieve a higher level of quality.

Quantity Measures

Measures of quantity represent a count of the amount of service or product delivered. As with quality measures, your organization may have an expected level of quantity that is associated with the resources you allocate to an effort or activity. If we return to our discussion of logic models and their use to describe the key components of a program, quantity measures specify the amount of outputs of a program that has been implemented. Your program funders are interested in the frequency at which your services have been provided and the number of individuals receiving services; they use this information as a means to monitor program activity. Again, your logic model articulates the connection between your outputs and outcomes of programmatic efforts. If your implementation of a program falls short with the delivery of its services or products, then the desired outcomes may not be achieved.

In the assessment of your organization's performance, if the quantity of services or products is not at the expected level, then you must determine if your organization's current processes impact your delivery of services or products. If your current processes and the allocation of resources need to be changed to be on target with the expected quantity of outputs, this represents an area for improvement. It is also important to keep in mind that in some situations you may have a higher quantity of services or products delivered than is needed. If so, this represents use of resources that could be allocated, more appropriately to another activity.

DEVELOPING YOUR PERFORMANCE MEASUREMENT SYSTEM

Per our earlier discussion, a performance measurement system acts like a barometer of how well your organization is operating and whether or not your efforts are achieving the results, as intended. When developing your performance measurement system there are several key steps, including (1) defining your purpose for measurement, (2) identifying what to measure, (3) specifying indicators, (4) developing a data collection plan, and (5) implementing a method to inform decision making.

Defining Your Purpose for Measurement

It is essential to first define the purpose(s) of your organization's performance measurement system and the intended audience(s) for the data gathered. Knowing this will provide guidance for developing your system with respect to (1) who gathers, analyzes, and reports the data; (2) what measures are used; and (3) at what times the data are

> *We must accept finite disappointment, but we must never lose infinite hope.*
>
> — Martin Luther King, Jr.

gathered and reported. To facilitate your understanding of the purpose of measurement, ask questions. For example, are your funders interested in both efficiency and effectiveness of your program for accountability purposes? Or is your management interested in efficiency of operations because you are faced with a situation of having to do more with the same?

Identifying Measurement Items

This component of your measurement system involves identifying the specific concepts to measure and track over time, based on the decisions made about purpose and audience. If efficiency is a focus as it relates to your organization's performance, then you must identify the specific items of interest, e.g., how long it takes to complete a process or specific steps in a process; how many staff members are involved in the process, and so forth. For example, you may have an intake process in your organization with an established standard that applications for service must be entered into your database system within three days of receipt of a completed application. To track this information, you need to set up a system that documents the date when an application is received, if the application is complete, and the date when the data from a completed application is entered into your database.

Specifying Indicators

For concepts and specific items you want to measure, you may have multiple indicators and different sources of data. Therefore, it is critical that your measurement system be clear about the specific indicators that will be used. For example, if you are measuring efficiency and one of your measures is the length of time it takes to complete a process, you need to determine if you are using minutes, days, weeks, and so forth to measure length of time. If you are measuring the effectiveness of your program, once you have identified your program's outcomes you must specify the indicators you will use to measure those outcomes. For example, if you are implementing a program to improve the level of engagement of students in a learning environment, you must identify the indicators of engagement, such as participation in discussion, hand-raising, eye contact, voice inflection, and the like, all of which are behaviors that can be observed in a classroom. You must also specify the length of time for the observation and if you will be doing a count of these behaviors. Alternatively, you may have an observer just rate the level of engagement for the entire classroom on a scale of 1 to 10, with 10 indicating a high level of engagement and 1 indicating a low level of engagement. In this case, the indicator used is more global in nature and based on the observer's subjective assessment.

Developing a Data Collection Plan

As with any system and its implementation, it is best to start with a plan. See the data collection plan in Example 4.1 for a quality improvement program for early care and education. Your plan will provide details about how data will be gathered, who will collect the data, how often data will be collected, and how data will be summarized and reported. For example, if you are measuring the amount of time (in minutes) it takes to complete a process step, your data collection plan should include information about (1) who will observe x number of staff completing the process step several times over x minutes to determine the amount of time it takes to complete the step each time, (2) how measurement indicators will be calculated, and (3) how these data will be presented in the report.

EXAMPLE 4.1. Data Collection Plan for ABC's Program to Improve the Quality of Early Care and Education

Agency	ABC Agency: Quality Improvement Program for Early Care and Education (ECE)
Mission	To create a quality improvement system in which all early learning programs and providers are encouraged and supported to improve child outcomes. Improvements in programming are designed to increase the capacity to support children's learning and development, increase educational attainment among practitioners, and enhance professional skills and competencies in support of children's learning and development.
Purpose of Measurement System	• To assess the efficiency of operations per the guidelines of program funders • To assess the effectiveness of program services in accomplishing the specified outcomes for early learning programs and providers

Efficiency Measurements				
Measure	**Indicator**	**Data Collection**	**Frequency**	**Reporting**
The time needed to process an application for a support grant and for the ECE program to receive the award.	The number of days from when a completed application is received to when the award payment is received by the ECE program. The target is ten days for 100 percent of the completed grant applications.	The grants manager enters into the database management system (1) the date completed application is received and (2) the date when a check is mailed to the ECE program.	The measure is reported at the end of each month for the activity during the month.	A bar chart is prepared that shows the number of completed grant applications processed, the range, and the average number of days from receipt to award for applications for that month, and the percentage processed in ten days or fewer.

Etc.				

Effectiveness Measurements				
Quality of classroom environment	The overall score on the environment rating scale, from 1 to 7, where 1 = low quality and 7 = high quality. Target is ≥4.5 average score for 80 percent of programs.	Reliable assessors observe one randomly selected classroom per grade level per ECE program. Scores on the subscale items are recorded in the database for each classroom observed.	One time per year.	A table is prepared that shows the number of ECE programs assessed, the range, average and standard deviation of the overall scores, and the percentage of programs with a score ≥ 4.5.

Etc.				

Implementing a Method to Inform Decision Making

While it is essential to gather and report the results of your data collection efforts, without a specific mechanism in place to use that data to inform decision making the tasks carried out to gather and report the data are WAs. As reiterated in the various models of continuous quality improvement and organizational learning, these data are to be used to inform the need for performance improvements. For example, if the amount of time it takes to complete your process (i.e., the sum of the times that it takes to complete all the steps within a process) is longer than expected, then decision makers need to use these data as a red flag to indicate that changes must be implemented to improve performance.

This requires a systematic process to gather, report, and review the data collected. The review process may vary from one organization to another and for individual performance measures. In some cases, data are used only by those implementing a process, and those data are used immediately to take corrective actions to improve the performance. For some other performance data, however, it will be necessary to have a process improvement team review the data and determine the next steps for improving performance.

RULES OF THUMB FOR DEVELOPING YOUR PERFORMANCE MEASUREMENT SYSTEM

Organizations are all different. Because of their unique circumstances, the purpose and intended audience(s) for reporting on performance will differ. One of the challenges in service organizations is the lack of well-established benchmarks that can offer guidance with respect to performance levels for many organizational operations. Furthermore, because individuals in service organizations often multitask when they work, it is difficult to capture data related to how long it takes to complete a process; those data are a key efficiency measure. These challenges and how to deal with them will be discussed further in Chapter 8. Regardless, there are a number of important rules of thumb to remember in the development of your performance measurement system; these rules of thumb are discussed below (Cunningham & Fiume, 2003).

Support Your Organization's Strategy

Your performance measurement system must be designed in such a way that it supports your organization's overall strategy and be structured to motivate the right behavior. It is often said that you get what you measure. Taking that into consideration, it is critical to select performance measures that are consistent with the behavior you want to encourage. Also, you should not measure that which you are not willing (or able) to change. Moreover, performance measures can have unintended consequences for other parts of the system, if those measures are driving the wrong behavior. It is human nature for people to find ways to make measures appear to be improving if they are being judged based on those measures. Take an organization that has a mission to improve the quality of care within childcare facilities. If this organization has implemented a program of education and technical assistance to improve the knowledge and skill of caregivers, then a key performance measure should be the number and percentage of caregivers that have improved their quality of care from before to after their participation in the program. There may be multiple ways to measure the quality of care, one of which is an environmental rating scale, a standardized observational instrument that assesses the extent to which an early care and education classroom meets research-based standards of quality related to space and furnishings, personal care routines, language-reasoning, learning activities, teacher–child and child–child interactions, program structure, and accommodations for parents and staff. Scores for each of these subscales range from 1 = inadequate to 7 = excellent.

Conversely, if the funder (e.g., a state agency) of the educational and technical assistance program has established targets as to the number of childcare facilities participating in the program, then the behavior of educational and technical assistance staff will be on getting childcare facilities enrolled so they meet their target. While this measurement will be much easier to collect and document over time, using this as a performance measure will be driving the wrong behavior—that is, getting enrollees rather than focusing on what needs to be done to ensure the enrollees improve their knowledge and skills. To minimize the impact of unintended consequences of a measure, additional measures must be used to provide a comprehensive assessment of organizational performance.

Be Few in Number, Simple, and Easy to Understand

At the same time, having too many performance measures will result in your staff spending all of their time collecting and analyzing data rather than doing the work they are supposed to be doing within your organization. Indeed, a few measures in the hands of many is better than many measures in the hands of few, since measures provide critical knowledge about performance to those within your organization that can make a difference and implement improvements. Moreover, since measures may be developed at various levels within your organization, depending on your organization's size, it is critical that those measures at the lowest level be ones that the line staff can relate to with respect to their own behavior. This is particularly important if your organization wants to establish a culture of continuous improvement and empower all workers to find ways to improve processes at their own level within your organization. Also, if the measurement system is complicated and difficult to implement, and if the resulting measures are not easily interpreted, the likelihood that the data will not be gathered routinely and that there will be confusion about what is really important in your organization is increased.

Be Mostly Nonfinancial Measures

While your organization is probably concerned about your bottom line whether you are for-profit or nonprofit, quality improvement within your organization should be about changes in processes to make those processes more efficient and effective. Your performance measurement system should consist primarily of measures that assess processes and how well they are accomplishing intended purposes. Ultimately, your performance measurement system will help to identify areas of waste so this waste can be eliminated. Resources used on WAs add to the cost of providing services. If waste is eliminated within your processes, the resources that are freed up may represent an improved bottom line or may be reallocated to other activities that are needed elsewhere for improvement in your overall organizational functioning. Therefore, financial measures must be used carefully, since process improvement is not just about saving money and increasing your bottom line.

Measure Process, Not People

While as a service organization you are mostly dependent on people (rather than machines) to complete work, if your organization wants to create a culture of continuous quality improvement, it is essential for you to focus on processes in place and whether they contribute to or inhibit the achievement of your organizational goals. Moreover, the individuals within your organization responsible for work activities need to feel empowered to make changes in processes when needed to improve results. If the performance of individuals becomes the focus of attention in your measurement system, there is an increased likelihood of defensiveness, lack of cooperation and willingness to change, dishonesty or deceitfulness with respect to performance, and so forth.

Set Ambitious Goals, Measure Actual Results

The intent of having performance measures in the hands of many is to empower workers to do what they can do to make a process better. It is important to give your workers an understanding of how far they need to go, not just communicate to them that they need to improve. Furthermore, establishing ambitious goals will require workers to stretch and think outside the box rather than simply to implement minor improvements to make a process a little bit better. However, the goals still need to be reasonable, or the workers will feel powerless to do anything to reach the goal(s).

Do Not Combine Measures into a Single Index

If a performance measure is complex with a number of items included in a single index, it will be unclear to your workers what they need to do to improve the overall index. Therefore, it is important to keep separate measures, each with its own goals, so that it is easier for people to understand what they need to focus on in their improvement efforts.

Be Timely in Reporting

Depending on your organization and process being analyzed, what is timely will vary (e.g., monthly, weekly, daily, hourly). Regardless, if the purpose of having performance measures is for your management and workers to take corrective action to reach a goal, then this information has to be timely so the corrective action can be taken and the goal achieved. If information about the achievement of a goal comes only after the fact rather than at regular intervals throughout a process, then it is too late to make needed changes.

Show Trend Lines

Continuous improvement means just that—it is ongoing over time. Therefore, actual results should be shown for the trends over time, as well as for each intermittent goal that is set for a period.

Be Visual

With sporting events, there is a visual scoreboard that keeps everyone apprised of where each team stands on the important measures of performance. In a similar way, if your workers and management are to have critical information about their performances so they can make adjustments and improvements when needed, then this information has to be visual. Moreover, displaying this information will signal what is important within your organization. There are many ways to be visual in displaying your performance measures, using tables, graphs, and other types of pictorial representations. See Example 4.2 for some different options regarding visual displays of performance.

Together, these rules of thumb provide you with additional guidance as you design your performance measurement system. Undoubtedly, there is no single way to measure performance within your organization. What you develop depends on your purpose as an organization and the identification of measures that are most critical in helping you determine how successful you are in reaching your goals with respect to operational efficiency and effectiveness.

EXAMPLE 4.2. Visual Displays of Performance Measurements

Efficiency Measurement

Focus of process improvement: To reduce the number of days it takes to process a grant application to ten days or fewer.

Target: One hundred percent of completed applications will be processed in ten days or fewer.

Timeline for improvement: Process improvement was implemented in April.

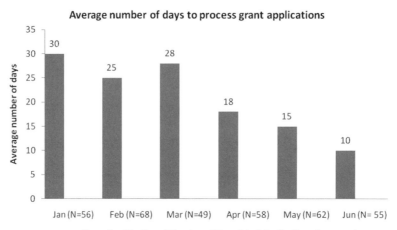

Average number of days to process grant applications

Percent of Applications Processed in 10 days or Less

Effectiveness Measurement

Focus of process improvement: To increase the percentage of programs that score ≥ 4.5 average score on the environment rating scale.

Target: Eighty percent of early childhood classrooms will have an average environment rating scale score of ≥ 4.5.

Timeline for improvement: Process improvement was implemented at the end of Quarter 2.

	Quarter 1	Quarter 2	Quarter 3	Quarter 4
Number of programs assessed	55	60	59	69
Range of scores	2.5-6.3	2.1-5.9	3.5-6.6	3.7-6.6
Average score	3.1	2.9	4.3	4.5
Standard deviation of scores	1.2	0.9	1.1	0.6
Percent of programs with ≥ 4.5 average	40%	38%	75%	80%

From Knowledge to Practice

Part II presents the key concepts of lean thinking as they have been adapted and made applicable to service organizations. The chapters in Part II also consider a number of the key concepts and methods that have been prevalent in understanding organizational behavior and quality improvement processes as discussed in Part I of this book. The following exercises provide you with the opportunity to take the knowledge gained in this section and put it into practice.

Exercise II.1. Creating a Service/Product Matrix

Exercise II.2. Identifying Wasteful Activities and Unacceptable Results

Exercise II.3. Establishing a Performance Measurement System

The final activity in Part II asks you to reflect on the exercises completed and discuss the challenges associated with completion of each exercise and any lessons learned as a result of your efforts that can help you in further attempts to apply the knowledge you have gained.

Exercise II.1. Creating a Service/Product Matrix

1. Create a service/product matrix for your organization or a division or department within your organization. Use Figure 3.3 as a model for doing this step.

2. Identify the factor you used to isolate the value streams in #1 above (i.e., your organization's client groups, the work volume of your organization's activities or programs, concerns voiced by your clients or customers, or specific business conditions that are critical to the success of your organization's work).

Exercise II.2. Identifying Wasteful Activities and Unacceptable Results

1. In the operation of one of your organization's programs, identify any of the following WAs that you or other stakeholders have experienced in the delivery of services or development and distribution of program materials (mark all that apply):

❑ Staff wait for missing information or data (from sources external or internal to the organization) that are needed to proceed with work.

❑ In some of the services provided or materials developed and delivered to our clients, there are mistakes made or built-in defects.

❑ It often takes too long to complete work or deliver a service to our clients or customers based on established expectations and timeframes for completion and delivery.

❑ Staff or clients do "rework" either because something was not done correctly the first time or because of the way a process is set up; duplicate work is being done.

❑ Inconsistencies or variations exist in the way the same work is completed either by different staff or by the same staff person across time.

❑ There are policies and procedures in our organization that require an excessive number of reviews and sign-offs to get approval to move forward on a proposed plan or project.

❑ Our management personnel spend an inordinate amount of time in meetings that seem to "go round and round" on issues, without making decisions that are needed for the organization.

❑ Our organizational office space is cluttered and disorganized, with materials piled in corners or stashed away in storage areas; no one is sure what materials are stored and how to efficiently retrieve specific items from storage, if needed.

❑ Our organization has a standard set of services or materials that we deliver to our clients; in some instances, these services and materials may not meet clients' needs.

❑ The workstations in our office are not arranged or set up for maximum efficiency of those staff that work at these stations.

❏ There are staff within our organization that always seem to have more work to complete than the amount of time available to do so.

2. For those areas of waste present in your organization that you identified above, provide a specific workplace example that demonstrates that waste.

Identify how any of these WAs produce URs in your organization and discuss how these URs fit within the FAQ-WOES categories, as discussed in Chapter 3 (e.g., Flow of work, Adequacy of resources, Quality of services or products, Workload balance, Organization of work environment, Effectiveness of process, Standardization of work).

Exercise II.3. Establishing a Performance Measurement System

Using the framework for specifying a performance measurement plan as shown in Example 4.1, prepare a perfor- mance measurement plan with at least two efficiency mea- sures and two effectiveness measures for a program your organization operates. Assume that the purpose of your measurement plan is to report to your funders, document- ing both the quantity and the quality of services provided, as well your accomplishment of program outcomes.

Reflections and Lessons Learned

1. Based on your completion of the exercises in Part II, reflect on both the challenges and lessons learned, and dis- cuss them.

2. Identify specific ways that you can improve your ability to use these lean concepts to specify value streams in your organizations, identify WAs within the value streams, and determine the impact these WAs have in terms of the URs for the efficiency and effectiveness of your organization's operations.

Using Lean Tools and Methods

With a basic understanding of the key concepts relevant to organizations engaging in lean transformations, we now turn to the lean tools and methods used to implement that transformation. In Chapters 5 through 7, we articulate how you can use the conceptual framework and apply the methodological tools to begin your lean journey. These tools include (1) understanding the current state of your organization's operational processes, (2) identifying the wasteful activities to be eliminated within your processes, (3) designing your future state where performance has been improved, and (4) implementing your planned improvements and tracking progress over time.

Chapter 5 returns to the value stream concept and provides an overview of the lean tools you can use to understand and visually represent a complete value stream and its subset of associated processes. These tools include value stream and process flow mapping. In addition, this chapter shows how to prepare for the mapping process by gathering data via surveys, key informant interviews, observation, and document review to identify those processes within your value streams that are problematic (i.e., that represent unacceptable results) to those that experience the processes. This chapter also discusses another preparatory step of establishing your core team of organizational staff and stakeholders to map the identified value stream and its processes.

Chapter 6 first discusses the tools you will use to analyze your value stream and process flow maps to determine the root cause of the wasteful activities and their unacceptable results; these WAs and URs are seen as problematic with respect to the optimal functioning of the process. The Five Whys and the Fishbone Diagram techniques are shown as very simple approaches to determine and understand root causes. In addition, this chapter offers an approach to identify the opportunities for improving processes and the priorities attached to these opportunities, keeping in mind that improvements can range from the very quick and simple to the complex, requiring considerable time and resources to implement.

The second part of Chapter 6 introduces a number of the potential solutions that you can implement in your organization as a means to ensure a successful lean transformation. The "5S" improvement tool is a critical first step, because it ensures work areas are systematically kept clean and organized, which establishes a foundation for building your lean organization. This chapter also discusses other tools used in lean transformations to solve operational problems, including workload balancing and the use of visual controls.

Chapter 7 provides an overview of several means to plan, communicate, and track the results of your lean transformation. The tools introduced in this chapter include the Action Planning Tool, a Success Stories table, Tracking Performance Measures graphs, and the A3 Report, an effective way to communicate the problem, analysis, corrective actions, and action plan on a single sheet.

Value Stream and Process Flow Mapping

As articulated in Chapter 3, a value stream includes all the steps or series of actions from beginning to end that are required for you to deliver a service or produce a deliverable for your clients. The end of the value stream is the point at which your clients receive the service or other deliverable. A VSM is a simple tool that shows how work is done and how to improve that work. Creating the VSM involves diagramming the macro steps of a process and specifying critical information about the flow of work and information through the value stream (Rother & Shook, 2003). The map also includes key data about the process steps, such as process times, number of people involved, and the service costs. The diagram, as a visual representation of a process, can be as simple as a paper and pencil drawing or a series of sticky notes that depict the major steps in a process.

TYPES AND LEVELS OF MAPS

Current State vs. Future State Maps

There are two types of maps. One type depicts the **current state** and visually represents the way information and workflow is occurring at the present time. The other type represents the **future state** and depicts the information and workflow after lean tools have been applied to eliminate WAs found in the current state. The process improvement plan is based on moving from the current state to the future state, as shown in Figure 5.1.

Value Stream vs. Process Flow Maps

Mapping also can be done at different levels to show the steps in completing a process to deliver a service or product to a client. One is a high-level view of a value stream (VSM); this view can be across a number of organizations, across a number of divisions within a single organization, or within a single division in an organization. Another is a lower-level view, where the subprocesses within a higher-level value stream are mapped. For our purposes, we use the term **process flow maps** for those that depict the subprocesses at this lower level.

FIGURE 5.1. Using the Value Stream Mapping Tool

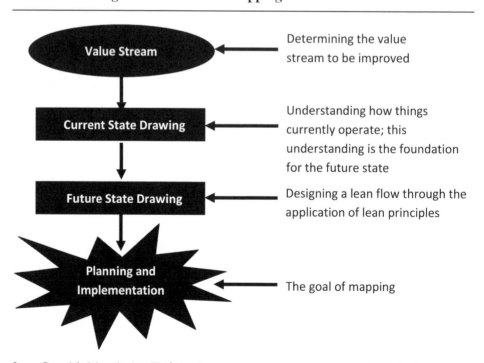

Source: From Administrative Lean™ of Lean Concepts, LLC. Retrieved April 24, 2009, from http://lean concepts.com/case_studies.htm

Key Features of Mapping in Lean Transformations

Regardless of the type of map and its level, there are a number of key features that differentiate the mapping process associated with lean transformations from other mapping techniques. When you do value stream and process flow mapping, you will do the following:

1. Use a systems perspective to identify the beginning and ending of the **value stream**; this value stream may cross over a number of functional areas either within or across your organization and other organizations.
2. Place a focus on **client needs and demands** when identifying the purpose of your targeted value stream.
3. Link the **flow of information**, initiated when your organization receives a signal that there is a client need for your service or product, to the actions implemented by your organization to meet that need.
4. Document key **performance measures** relevant to the delivery time and quality of your service or product.

Mapping the value stream and its subprocesses is a critical first step when your organization begins its lean journey. These maps provide the foundation for your lean transformation, because they will help you

1. gain an understanding of how—and how well—your process works;
2. see where WAs and URs occur; these factors impact the flow of work through your value stream;
3. reach agreements on what changes need to be made to improve your process and organizational performance; and
4. reach agreements on how to ensure the changes are made to improve processes and performance.

PREPARING FOR THE MAPPING PROCESS

Before your mapping process begins, there are some preparations you must make. Undoubtedly, there are a vast number of work processes that take place at different levels within your organization or system structure. This process complexity raises a number of questions as it relates to process improvement efforts within your organization. The initial questions you must address are these:

1. What is the driving need for your process improvement efforts?
2. How do you know what process to map?
3. What level of detail should you depict in your mapping?
4. Who should be part of a process improvement team effort?
5. What operational performance and outcome measures should you use?

Given these questions, your preparations include

1. gathering data via surveys, and, possibly, via key informant interviews, observation, and document review to identify processes that are problematic to those that experience them;
2. establishing a core team of organizational staff and stakeholders to map and analyze the identified value stream and it processes; and
3. specifying key performance measures for the processes and organizational outcome measures.

Gathering Data about Problem Areas

Ideally, your organization has established a culture where there is continuous and systematic examination of your processes with respect to their efficiency and effectiveness. Your managers identify performance concerns as part of their oversight responsibilities; this oversight might be based on information flowing from the internal staff or from clients. Organizational inertia is more the norm, however: when times are good, you may be less likely to see the need to change or improve your current pattern of behavior. In times of crisis, though, your management team receives a message that it is necessary to take a critical look at operations; this situation provides you an opportunity to create a lean organization.

Regardless of the driver for implementing a change to improve your organizational performance, the initial step of bringing focus to the improvement effort involves gathering input with respect to how well your organization's current processes are working and where URs are occurring, within either a process or a set of processes across a value stream. This input provides evidence that there is a need for applying lean tools to transform your organization and improve performance. The ways for you to gather this input include the use of surveys, interviews, observations, and reviews of documents or records.

A Survey of Targeted Stakeholders. Returning to our discussion of WAs and URs, you can create a survey tool to gather opinions of targeted stakeholders (e.g., line staff that implement one or more steps of your process, internal or external clients of your service or other deliverable, executive and management staff, and funders). Form 5.1 in the appendix at the end of this chapter provides a format for this type of survey to determine the extent to which your organization is impacted by WAs. A form similar to this can be developed for determining the impact of URs, as shown in Form 5.2 in the appendix.

You can analyze the responses to your survey and create a rank order of the various types of WAs and URs. (See Table 5.1 for the types of analysis and how to report results of analysis based on three different ways to rank order responses with respect to WAs.) Furthermore, you can determine the particular set of processes where these issues have a significant impact through an analysis of open-ended comments that respondents make. This survey assesses the extent to which your organization has current operational processes that can benefit from further application of the lean tools to bring about significant improvements in performance. In addition, your analysis of the responses to this survey can include a breakdown of the opinions by type of respondent to determine if there are significant differences among them with respect to how they experience a process.

Key Informant Interviews. In addition to a large, broad-based survey as described above, another source of information about your organizational performance and the efficiency and effectiveness of your processes is through key informant interviews. Depending on the results from your initial survey, key informant interviews may not be conducted during this preparatory stage, particularly if sufficient input was gathered from the survey to provide a focus for your improvement efforts. If that is the case, the key informant interviews may be done after the improvement team begins the mapping process and adds the critical data to the value stream and process flow maps. The purpose of these interviews is to gain a greater depth of knowledge about how your process works, along with insight as to the areas of waste and URs associated with the process you have identified as problematic. The key informants will be a critical source of information about these details, and should consist of individual representatives of

TABLE 5.1. Wasteful Activities Survey Results Sorted by Mean, Very Significant Impact, and Significant Impact

Waste (sorted by mean value)

Waste is found in processes (N* = 61)	N	Min	Max	Mean	Std. Deviation
Resource depletion	60	1.00	5.00	3.20	1.23
Waiting	61	1.00	5.00	3.15	1.30
Extra processing steps	61	1.00	5.00	2.95	1.06
Convoluted pathways	61	1.00	5.00	2.95	1.12
Information deficits	60	1.00	5.00	2.93	1.19
Process variation	61	1.00	5.00	2.90	1.01
Required rework	59	1.00	5.00	2.80	1.13
Errors or defects in services or materials	59	1.00	5.00	2.75	1.14
Inefficient work stations	60	1.00	5.00	2.63	1.22
Stockpiling materials and supplies	61	1.00	5.00	2.39	1.17
Delivering more services or materials than needed	61	1.00	5.00	2.26	1.14

Waste (sorted by "very significant impact")

Waste is found in processes from (N = 62)	No Impact	Some Impact	Neutral	Significant Impact	Very Significant Impact	Missing Data
Waiting	8%	34%	10%	29%	18%	1%
Resource depletion	10%	22%	16%	36%	13%	3%
Convoluted pathways	5%	40%	16%	29%	8%	1%
Inefficient work stations	18%	34%	19%	18%	8%	3%
Information deficits	11%	31%	14%	34%	7%	3%
Required rework	6%	45%	11%	26%	7%	5%
Stockpiling materials and supplies	22%	40%	16%	13%	7%	2%
Extra processing steps	5%	37%	19%	32%	5%	2%
Errors or defects in services or materials	16%	24%	26%	26%	3%	5%
Delivering more services or materials than needed	32%	27%	21%	16%	2%	2%
Process variation	5%	39%	18%	36%	1%	1%

Waste (sorted by "very significant and significant impact")

Waste is found in processes from (N = 62)	No Impact	Some Impact	Neutral	Significant and Very Significant Impact	Missing Data
Resource depletion	10%	22%	16%	49%	3%
Waiting	8%	34%	10%	47%	1%
Information deficits	11%	31%	14%	41%	3%
Extra processing steps	5%	37%	19%	37%	2%
Process variation	5%	39%	18%	37%	1%
Convoluted pathways	5%	40%	16%	37%	1%
Required rework	6%	45%	11%	33%	5%
Errors or defects in services or materials	16%	24%	26%	29%	5%
Inefficient work stations	18%	34%	19%	26%	3%
Stockpiling materials and supplies	22%	40%	16%	20%	2%
Delivering more services or materials than needed	32%	27%	21%	18%	2%

N = number of respondents

different stakeholder groups who have a perspective to offer as it relates to your targeted processes or value stream.

Observation of Work Processes. Along with the input from your survey and key informant interviews, observation is another method to gather information about how well your process is implemented. As with the key informant interviews, observation may be done when your improvement team begins the mapping and analysis work. Depending on the timing of your observation of a process in action, the observational tool will have a different structure and purpose. If observation is done prior to establishing a specific value stream and subset of processes to map and analyze, then your observational tool will be similar to the survey used across stakeholders to solicit input on problems associated with your organizational process(es). The focus of your observation will be wider and cover a broader spectrum of your organization's operations. However, if observation is done to gather information about a specific process as it relates to the key data elements that must be attached to a VSM, then the observational tool and the methodology for completing your observations will be different. We will provide more details on this later in this chapter in our discussion of the data elements attached to VSMs.

Examination of Critical Documents and Records. When your targeted process utilizes either documents or records (e.g., application forms, hard copy or electronic informational records, and so forth) they need to be identified and gathered together; they will become a critical piece of information once mapping begins. However, if your initial survey results highlight problems associated with forms and the ways your organization communicates and moves information through it, then these documents and records should be gathered upfront and used by your improvement team as they begin their work to map, analyze, and improve your process.

Other Data-Gathering Rules of Thumb

Other decisions that have to be made during this initial step of gathering input are with respect to *how many* and *who* the particular stakeholders and key informants are. These decisions will vary, depending on the particular circumstances within your organization, but a rule of thumb is that the survey group will be larger than the number of key informants interviewed. Surveys of this nature can be easily administered via the Internet; that kind of survey will eliminate a data entry step because the responses are captured in a database that can be readily imported into any data analysis software program where the results can be tabulated and further analyzed with respect to rank order.

The key informants should be a subset of your survey group, selecting those individuals that represent each stakeholder group from which input is being gathered. In addition, the key informants should be selected based on their depth of knowledge about the particular processes that are likely to be the focus of improvement efforts.

There are some additional rules of thumb to use when embarking on a lean transformation within your organization. First, it is important that you focus your improvement efforts on *repetitive processes*. If a work process occurred only as a unique response to a particular situation, then it is a rare occurrence and should not be the focus of your improvement efforts since you will not have a return on your investment of time and energy to improve the process. Work processes that are routine and repetitive within your organization are those that should be mapped to identify improvement opportunities.

In addition, you can use these other criteria to select a work process on which to focus your process improvement efforts: (1) select a work process that has the **most client complaints**; (2) select a work process that results in repetitive **quality defects or errors**, or (3) select a work process that has significant **cost variation** between what is planned and actual.

Furthermore, for the most effective mapping exercise, it is best to look at a process where you have a **discrete set of tasks** designed to deliver a service or produce a deliverable, and information is flowing from person to person. When mapping is done at too high a level, it often is vague and does not yield results with concrete applications.

In addition, it is best to do **mapping in the physical place** where the process is implemented. This is more practical, in that you will have the staff, equipment, and procedural rules connected with your process close at hand, enabling you to construct a process flow map more easily.

Establishing a Core Team

Establishing the team of individuals that will complete your mapping, analysis, and process improvement efforts is another task to complete prior to beginning your lean journey. In establishing the team, it is important to remember a number of key elements for effective teaming, without which your process improvement effort may be doomed. If the right people are not in attendance, then both time and money will be wasted. For example, if your team does not include someone in authority who can make decisions when decisions have to be made, your team may avoid making decisions for fear management will not agree, or may make decisions that are later overturned by management.

Therefore, when forming your team, the first task is to **clearly state the goals or objectives** for your team to be assembled based on organizational needs or problems that must be fixed, followed by **identifying the right people** so that those objectives can be accomplished. For example, your organization may need to justify budget and funding expenditures, ensure accountability, optimize use of resources, establish performance baselines, understand program effectiveness, improve turnaround time, or reduce costs. At this point, it is also important to identify who wants to know about the results of your improvement effort and whether these individuals are internal or external to the organization.

At the beginning stage of your process improvement effort, the objectives may be somewhat general if there has not been any data collection about process and performance issues: these data will provide more focus to the effort. Therefore, the actual team membership may change over time as your focus and objectives become more specific. Regardless, the following provides some guidelines in determining who your team members should be:

1. **A team champion** who has the authority to make decisions and dedicate resources to the process improvement effort. The champion also has the primary responsibility to remove obstacles, mentor and support the team, verify that a project is aligned with an organization's mission or goals, keep projects focused, ensure timely completion, and drive quantifiable results.

2. **A process owner** who is a senior manager or department head and has management responsibility over and a rich experience with the targeted problematic process or functional area.

3. **An internal or external expert in lean process improvement tools** who can provide expert facilitation for your team as it progresses through the steps of identifying a performance issue that can be addressed through the use of lean concepts and methods.

4. **Several staff members who are close to the process(es)** to be mapped and analyzed. These staff are **subject matter experts** who will have the depth of knowledge and factual information regarding how and how well a process works; it can also include staff who can retrieve detailed cost data associated with the operation of a process.

5. **One or more staff within your organization that are not part of the value stream** and its processes being mapped, analyzed, and improved. These staff provide an outsider's view of the process(es) under scrutiny and are able to ask the types of questions that help to clarify how a process works and why various steps in a process are performed in a particular way.

6. Ideally, **one or more clients** that are the recipients of your service or product delivered by the organization. Clients, of course, have the inside knowledge about what is of value to them, an essential piece of knowledge that your team will need once it begins the determination of whether a particular activity within a process adds value to what is delivered to your clients and whether your clients actually need or want the service or product of your organization. Keep in mind that, depending on the process that is the focus of your improvement effort, there may be both internal and external clients that can make up the team.

7. Ideally, **one or more other stakeholders** that have a connection with the process being examined (e.g., funders, board members, and so forth). As with the clients and team members from another part of your organization, the stakeholder representative(s) can provide an outsider's perspective that is invaluable when mapping and analyzing processes.

> *We find comfort among those who agree with us— growth among those who don't.*
> — Frank A. Clark

In addition to the types of team members as discussed above, consider the **size of the team**. There is no single one-size-fits-all when it comes to the number of team members. Determining the size of your team may be driven by the size of your organization, the particular process that will be mapped, analyzed, and improved, and the availability of internal staff, as well as the availability of other stakeholders who should be on the team. However, teams that are too large become cumbersome to facilitate effectively; a rule of thumb is that teams should not be any larger fifteen individuals.

Furthermore, some team members will only be needed at certain points in your process improvement efforts, and the whole team only comes together for status reports. For example, a team champion or some other upper-level management personnel may only be at the kick-off meeting and regularly thereafter be updated with status reports. Furthermore, there may be times when subgroups of your team will have separate tasks to complete that are done outside of any team meetings.

Specifying Performance Measures

One of the most important elements of a successful process improvement effort is to have specific types of measure to track the results of changes made. As discussed in Chapter 4, there are different types of performance measures that provide evidence of your organization's efficiency or effectiveness, as well as information about the quantity and quality of services or products provided. While all these types of measures are important in one way or another with respect to tracking performance over time, some of these measures are best used to provide evidence of the long-term impact of your organization's activities rather than to provide a measure of how well your organization implements its processes. For example, in many service organizations it is critical to assess if programs are effective in accomplishing their mission or long-term goals. Thus, if a four-year-old educational program has a mission of ensuring that its students meet or exceed a designated level of knowledge and skill, then the knowledge and skill outcome measures will be gathered prior to and at the conclusion of the educational program. These are measures of the program's effectiveness in accomplishing its stated goal.

For the purposes of improving specific organizational operations associated with a value stream or specific processes within a value stream, it is vital that you have performance measures that give you frequent information that is easy to gather about whether your planned and implemented improvements are producing the expected results. Without this immediate feedback, your ability to quickly adapt and make improvements is hampered. Therefore, the specific performance measures relevant at this point are those that assess the short-term results of your implemented changes to improve the operation of your process (i.e., to improve its efficiency or the quality of the service or product delivered to your clients). Examples of process measures most relevant to service organizations include processing time, batch sizes or practices, demand rate, percent complete and accurate, reliability, number of people, inventory, information technology used, and available time (Keyte & Locher, 2004).

Processing Time. There are several time-related measures important in lean thinking, including the amount of time elapsed from the start to the end of a process (elapsed time), the amount of time it takes to complete a step of activity within a process (cycle time), the portion of time within the cycle time that adds value with respect to meeting client needs (VA time), and the amount of time spent waiting between process steps (wait time). Also, changeover time may be relevant in some of your value streams because it represents the amount of time it takes to change from one activity to another. This may happen in your service environment if your workers are interrupted frequently when doing their work or must divert their attention to another work activity that is required to complete the initial process activity, e.g., the need to work around the process due to information deficits, atypical situations, and so on.

Batch Sizes or Practices. This measure represents how much or how often work is performed. For example, if your organization has a routine of only doing certain types of work on a schedule, such as once a week, then the batch size is one week's worth of work.

Demand Rate. Demand rate represents the volume of transactions over a specified period (e.g., number of applications per day that are processed). This is an important attribute in that it can be related to your client needs or demands. Demand rate is important information in designing a system to meet a specific demand rate.

Percent Complete and Accurate. This measure is a quality measure that describes how often a downstream process step receives material or information that is complete and accurate so that the next activity can be completed (e.g., all the information in an application is there and legible, allowing your worker to begin processing the application).

Reliability. Reliability is the percentage of time that a piece of equipment is available in your organization when needed (e.g., a completely functional computer and its software programs, copy machine, fax machine, etc.).

Number of People. This measure is the number of people trained to do or who are responsible for doing specific work in your organization. You may have more people trained to do the work although the work is actually completed by fewer people. This number is an indicator of how widespread the task knowledge is in your organization, which may raise issues as to the amount of cross-training, or lack thereof.

Inventory. Inventory can represent two different attributes of a process. The first is the size of a queue holding information or materials waiting to be processed. This type of queue is associated with batch processing and a long elapsed time for a process to be completed. Another use of the inventory measure specifies the amount of material that is stockpiled, waiting to be used in a process. This may include supplies and equipment that sit idle and consume your organization's resources (e.g., dollars for their purchase, space for their storage, and personnel time to manage the stockpile).

Information Technology Used. This measure describes the information technology (software) used in a process. This is not a true performance measure, but it is an attribute of a process that should be captured because information technology systems within your organization can either inhibit or facilitate processing time.

Available Time. This measure is the amount of time that your organization or individual worker has to do actual work, thus it represents the effective working time over the course of a day. A rule of thumb for determining effective working time is to use fifty minutes for each hour a worker is available to complete work, and to take lunch and breaks out of the total hours your workers are at work. Therefore, in an eight-hour day (480 minutes), if you take away thirty minutes for lunch and twenty minutes for two ten-minute breaks, along with another eighty minutes of noneffective work time, the time remaining (350 minutes) is the available time in a worker's day to complete work tasks. This measure is important in that it helps you determine the necessary capacity required to meet a specific demand. For example, if it takes thirty-five minutes to process an application for a service and an organization receives fifty applications in a day, then a single worker can complete ten applications in a day's time. If your organization's goal is to process an application within a day, then you will need five staff to process applications to meet that demand.

Determining the performance measures to use when mapping a value stream can appear to be a daunting task, particularly if you are a service organization that has collected little performance data. If so, it is important to take the time to determine the performance measures that will drive the right behavior with respect to operational efficiency. You must ensure that performance measures are those that also will support and enhance the effectiveness of your processes. Moreover, these performance measures must be quick and easy to document and collected on an ongoing basis to provide feedback with respect to the results of process improvement efforts. With this in mind, the following are keys points to remember in the selection of process performance measures (Keyte & Locher, 2004):

1. The purpose of the measures is to help you visualize the operation of a process and to identify process issues. Therefore, while the performance measures discussed above offer a set of measures often used in lean transformations (where cost, quality, and customer service are most important), they are not intended to

represent every situation. Depending on the particular value stream mapped and the presenting issue with respect to the value stream, the measures used need to be those that will *tell the story* by making the operation of your process visual.

2. The measures should be those that highlight areas of waste impediments to the flow of work, and the URs of processes that are experienced by your staff, clients, or other stakeholders.

To help in the identification of the measures that are relevant to a particular value stream mapping, analysis, and process improvement project, there are a set of questions that your process improvement team needs to ask. These questions provide the foundation for implementing changes and tracking the results of these changes to determine if the desired result(s) occurred. While there are a number of performance measures relevant to lean thinking, as discussed above, in the preparation for a process improvement project the performance measures specified as a response to the questions below are those that will be tracked to determine the success of the improvement efforts. These four questions are

1. What is the purpose of the process?
2. What are the indicators of success?
3. What are the specific performance measures?
4. Who is responsible for the process?

For example, let's take a work process that is quite common in many human service organizations—the preparation and distribution of a newsletter to a set of individuals that are the organization's network, and to other stakeholders within their network. See Table 5.2 for how these four questions were answered for this example. Once the improvement team has agreed on answers to these four questions, the team can move forward with the mapping of the process.

TABLE 5.2. Questions and Answers in Preparation for Mapping a Work Process

Questions:	Answers:
1. What is the purpose of the process?	To develop and distribute a newsletter to the organization's clients and stakeholders in order to meet their need for information regarding the program and its delivery of services.
2. What are the indicators of success?	• The newsletter is mailed on the specified due date. • The newsletter preparation process takes ten working days to complete. • The newsletter has no errors in it.
3. What are the performance measures?	• Number of days before or after (+ or –) the specified date for mailing. • Number of days (+ or –) to produce the newsletter compared to the expected number of days. • Number of errors within the newsletter.
4. Who is responsible for the process?	The administrative assistant manages the process and coordinates the activities of the newsletter contributors, the printer, and the mailing room staff.

CREATING CURRENT STATE MAPS

As previously discussed, VSMs identify the macrolevel steps in a process, including the identification of both value and non-value added activities. VSMs are the blueprints for lean transformations. A VSM includes every major action in the flow of material and information, including key data (Womack & Jones, 2003).

Key Performance Data in Value Stream Maps

In lean transformations, the key data related to time, labor, and materials are critical in the analysis of the value stream for areas of waste that can be eliminated, along with areas where other lean tools can be applied to make process improvements. Key data that are specified in your VSM may include any number of process measures that are relevant to the process being mapped, as previously described. The following list includes those six measures that are essential when mapping a process to eliminate waste and non-valued-added activities. Four of these process performance measures are time related, and the other two provide information about the resources dedicated to the operation of a process, i.e., the number of people performing the process and the monetary value of the supplies or materials consumed during the process:

1. The **total cycle time** (CT) it takes to complete a step or activity in the process.
2. The **time that is value-added** (VAT) as a portion of cycle time.
3. The **number of individuals involved** (J) in a process step.
4. The **amount of waiting time** (WT) between the steps within the process.
5. The **elapsed time** (ET) from beginning to end (i.e., CT + WT).
6. If relevant, the **dollar value** ($) of the supplies or materials in work or in stock (inventory).

In your service organization there may be challenges to specifying these key pieces of data, because the nature of your work is different from that done on a factory floor. First, humans do the majority of your work activities to deliver a service. Often they are doing multiple tasks within any specified period—and are not just focusing on activities associated with one process within a value stream. Therefore, interruptions may occur in the completion of a process step. Also, there may be variations in worker capability, the type of client, or the type of item being processed. Given these variables, it is not easy to identify the time-related data for completing specific tasks because of the degree of process variation, both when a single person completes a task and when a number of individuals complete the same tasks.

Second, for some types of activities associated with delivering services, you may not have any information or benchmarks with respect to how other similar service organizations operate. As a result, it is difficult to gauge client demand or needs, or what is best practice with respect to the operation of a process. For example, the governmental agency that awarded grants to organizations for economic development projects—see Example 1.1—did not have information about what is a realistic amount of time it should take to process an application for a grant. Moreover, based on current processes, the amount of time to process each grant application might be different, since some applications may be complete when they are submitted, while others may be missing information that needs to be gathered before processing can begin.

Given these challenges to specifying the data regarding time, labor, and materials for a value stream or process that is mapped, we offer these recommendations as to how you should approach specifying these data:

1. Have those staff that do the work provide their best estimates of the required data that are sufficient for the purposes of creating the current state map—i.e., the time or range of time it takes to complete a task, the number of people involved, and the cost of materials. If your staff are uncertain about these data or are reluctant to estimate the time they spend on a task, have them gather the information about the number of people and cost of materials from the appropriate sources within the organization. For the time data, have staff perform the steps of a process over an appropriate length of time, which will vary depending on the process steps being completed, and to document the cycle time, VA time, wait time, and total elapsed time. If there are variations across workers or items being processed, then data can be displayed as a range of time (e.g., five to ten minutes).

2. Have someone other than the person responsible for the work complete a short observation of the work tasks, to gather sufficient information to calculate an average or range of times (e.g., CT, VAT, WT, and ET) associated with completing the steps of a process, the total number of people involved, and the amount of material consumed; the amount of material can have a cost attached based on unit costs.

Steps in Creating the Value Stream and Process Flow Maps

To illustrate the steps involved in creating a VSM, we have taken our example of the value stream for the creation and distribution of a newsletter. In the organization in this example, the prevailing UR is that it *takes too long* for the newsletter to be completed and distributed—often it gets into the hands of the clients and stakeholders at a point when the news are no longer new and deadline dates for certain activities have already passed.

With the UR in hand, the value stream to be mapped has already been identified, with a designated beginning and ending point (i.e., it begins when contributors are informed of the schedule for creating the newsletter and ends with the newsletter's distribution). Furthermore, the preparatory step of answering the four questions about purpose and performance measurements have been completed, as shown in Table 5.2. With these answers as a given, the following describes the steps involved in creating a VSM for our newsletter example.

Step 1: Identify the Macro Steps of the Value Stream. This VSM of the current state shows the macro steps in the work process to create and distribute the newsletter. The macro steps are displayed in a **dedicated process box**, as depicted in Figure 5.2a. The macro steps of this value stream include create, lay out, proof, print, assemble, and distribute. The *material* that is flowing through this process is the newsletter, with its intellectual content; the newsletter is being developed and prepared for bulk mailing as it travels through each of the macro steps. In this VSM, the newsletter is a tangible item that is produced and not a client moving through the value stream that receives a service from the organization, as is the case in many service organizations. The flow of the newsletter through the macro steps of the VSM is depicted with a **material or client flow arrow.**

> *Remember: today is the tomorrow you worried about yesterday.*
> — Dale Carnegie

A rule of thumb in identifying the macro steps of a value stream is to look at the system process from a higher level of abstraction and capture the separate *chunks* of activity that can be easily described with a single term. Remember that these macro steps are processes, not functional departments within an organization. If desired, the department responsible for a process can be noted in the process box, but it is not necessary.

FIGURE 5.2a. Value Stream Map of Newsletter Process: Macro Steps Within Dedicated Process Boxes

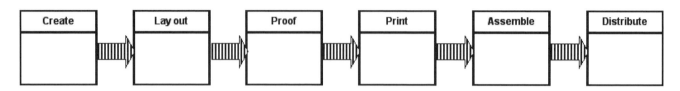

Step 2: Specify the Information Flow. The next item to include in this VSM is the information that is flowing through the process, information that triggers the actions expected of those individuals that are part of the work process. In the newsletter example that we are using, the schedule for the development and distribution of the newsletter represents the information flow, which is done electronically, as shown by the **jagged arrow.** If the flow of information is done manually (i.e., the schedule is on a piece of paper that is distributed to persons involved), it is represented by a **straight arrow.** Figure 5.2b shows the VSM with information flow depicted.

FIGURE 5.2b. Value Stream Map of Newsletter Process: Information Flow

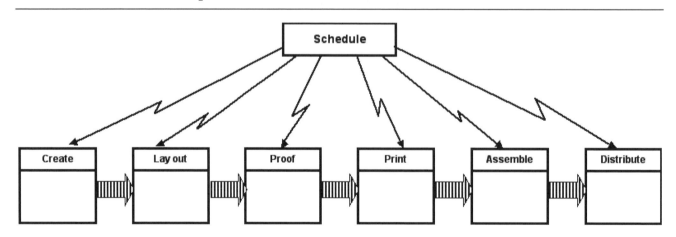

Step 3: Specify Selected Performance Measures. Using the identified performance measures to be documented for this value stream, include these data in the **attribute area box**, located beneath the dedicated process box. For the example of the newsletter value stream as shown in Figure 5.2c, the selected performance measures are (1) the **number of individuals** (☺) involved for each of the steps, (2) the **cycle time** (CT) for each process step; (3) the amount of cycle time that is **value-added time** (VAT), the total **elapsed time** (ET), and (4) the **wait time** (WT) between the macro steps of the process. In addition, the **dollar value** ($) of the materials consumed to produce the newsletter is specified under the process box where it is relevant.

FIGURE 5.2c. **Value Stream Map of Newsletter Process: Tabulated Data for Selected Performance Measures**

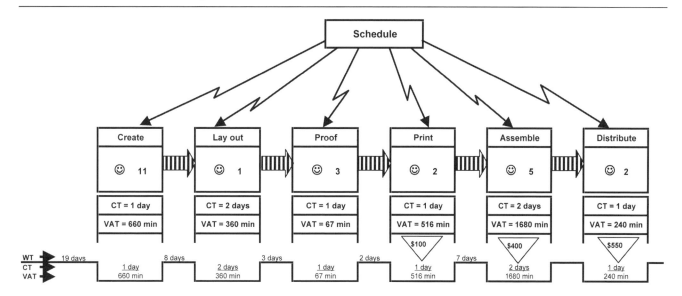

Total Number of Individuals Involved (☺) = 24
Total Cycle Time (CT) = 8 days
Total Value-Added Time (VAT) = 2.5 days = 58.7 hours = 3,523 min
Total Elapsed Time (ET) = 47 days = 376 hours = 22,560 min
Total Wait Time (WT) = 39 days
Total Value of Material (▽) = $1,050

Step 4: Perform Value Stream Walk-Through. As a means to fully grasp how a work is created, progresses, and is organized, it is important for the process improvement team to do a walk-through of the process, preferably starting at the end of the process and traveling upstream to each of the earlier steps of the process. This walk-through serves three key purposes: (1) to determine how clear the process steps are, (2) to show the process steps to others not familiar with it and to solicit questions from them to clarify the process, and (3) to introduce the concept of *pull* rather than *push* as the driver for initiating a process. While doing the walk-through, many of the performance measures should be collected (or a means to gather these data should be put in place, with the data added to the VSM at a later point). At this point, the team should be asking questions that will help them fully understand how work is done, how staff prioritize their work, and what issues are related to work. It is important to identify the work steps that add value to the service or product delivered to clients and where the areas of waste are. Keeping in mind lean philosophy, as discussed earlier, the team should be assessing whether a process step is (Womack & Jones, 2003)

1. Valuable: Does the step create value for the client?
2. Capable: Does the step produce a good result every time?
3. Available: Does the step produce the desired output, not just the desired quality, needed by the downstream (internal) client every time?
4. Adequate: Is the capacity for the step adequate so the client receiving the service or product being produced does not need to wait for the next step of the process?
5. Flexible: Can the step be switched quickly from one service or product to the next so services can be delivered or items produced in small lots?

Step 5: Create Process Flow Maps. You create a process flow map for the individual macro steps to understand fully the flow of work within a macro step and to estimate the amount of time that is VA (green dot), NVA (red dot), or RNVA (yellow dot). You make this determination as to whether a process step is VA, NVA, or RNVA as a team through discussion that occurs during the mapping process. As discussed previously, sometimes this determination does not come easily, particularly when members of your team are doing the work, since it is difficult to say you are engaged in NVA work. However, this discussion among team members is critical, since it raises questions about what does add value to the delivery of a service or product to a client.

Process flow mapping is another tool to represent visually how a process works, but the level of detail depicts a process with all the tasks, decision points, and flow of work based on previous steps and decisions. Table 5.3 provides the icons typically used for creating a process flow map.

However, because this mapping is not done for purposes of developing software programs or precise work instructions, absolute accuracy on these process flow maps is not required. Regardless, this level of detail for a macro step is valuable for identifying where there are points in a process that require a *work around* or some other NVA activity. *Work arounds* will happen when a downstream step in a process cannot be initiated until client or material items flowing through the process meet all the criteria for moving on to the next step. A good example of this is in your processing of applications for a service a client needs. If, upon receipt of an application, your initial review indicates that required data are missing, then your staff person must begin another process to gather the missing data before the application can move forward to determine eligibility of the applicant. For our VSM of the newsletter, a process flow map has been created for the *Create* macro step and is shown in Figure 5.3.

TABLE 5.3. Icons Used for Process Flow Mapping

Icon	Description
(rounded rectangle)	Start or end of a process
(rectangle)	Description of a task or process step
(diamond)	Decision point where work or information can flow in either direction depending on conditions
(arrow)	Direction of flow of work from one process step to another
(documents)	Documents used or created within a process
(cylinder)	Storage of information
(bracketed box)	Predefined standard for how a process works (e.g., government regulation)
(circle with A)	Connectors for a process that continues on another page or in another document

FIGURE 5.3. PFM for Newsletter: *Create* Macro Step

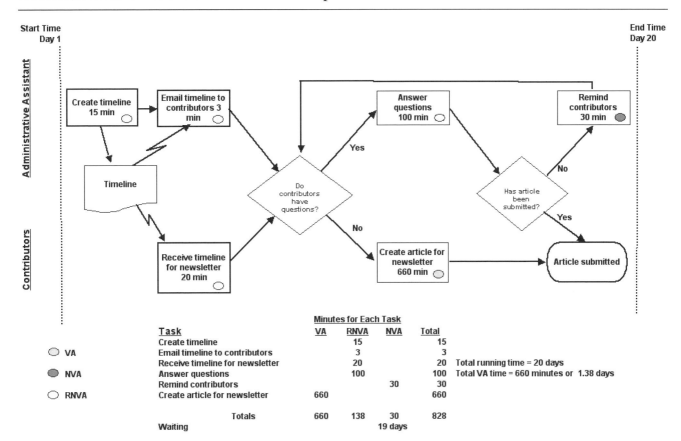

Once a process flow map is completed for a macro step, calculations must be made with respect to the percentage of time that is VA, NVA, or RNVA. For the *Create* macro step, these data are shown in Table 5. 4. (See the appendix, Form 5.3, for a blank copy of this form.) In Chapter 6, we will return to these data, since they are critical in identifying specific problems to be addressed in an effort to improve the process.

Step 6: Calculate Summary Performance Measures. Once the VSM and individual process flow maps have been created to visualize how a process works and the associated performance measures have been added to the maps, you can summarize the data to show the key information about time (total cycle time, amount of cycle time that is VA, total wait time, and total elapsed time), the number of individuals involved, and costs associated with materials and supplies. Also, you can note any additional performance measures that are a part of the summary measures. For example, if we return to Table 5.2 where the measures of success were identified for the newsletter process, the three additional performance measures are these: (1) the newsletter is mailed on the specified due date, (2) the newsletter preparation process takes ten working days to complete, and (3) the newsletter has no errors. The number of errors detected in the newsletter, once printed, is the indicator for the performance measure of *percent complete and accurate* as discussed earlier, while the first two measures are indicators of *processing time* that are specific to the newsletter process.

TABLE 5.4. Process Flow Map Data Collection: *Create* Macro Step

Macro Step: *Create*   **Elapsed Time (CT + WT) =**   **8000 minutes ***

* This represents the number of minutes in 20 days, with 8 hours per day and 50 effective minutes per hour.

Description of Subtask	Amount of Time*		
	VA	NVA	RNVA
Create timeline			15
Email timeline to contributors			3
Receive timeline for newsletter			20
Answer questions			100
Remind contributors		30	
Create article for newsletter	660		
Waiting		7600	
Totals	660	7630	138
Ratios of VA, NVA, and RNVA time to elapsed time:	8%	91%	2%

Specify the amount of time allocated to each subtask within the column that is appropriate, depending on whether the subtask is VA = value-added, NVA = non-value-added, or RNVA = required non-value-added.

The VSM provides some useful information about the current state, which helps us identify where to focus initial process improvement efforts. In this particular example, knowing that there are eleven individuals involved in the creation of content (for a four-page newsletter), and knowing that this front end of the process is where the bulk of waiting time exists, it seems appropriate to address process improvement efforts at this macro step of the work process. Specifically, we see that in the *Create* macro step, which takes twenty days, there are only 660 minutes spent in VA time, equal to 1.38 days out of the 20.

TIPS FOR MAPPING THE CURRENT STATE

The goal of creating current state value stream and process flow maps is to show clearly where WAs are occurring; these can be eliminated once further analysis has been completed to determine the root cause for issues that have been pinpointed in the current state maps. This next step for implementing a lean transformation will be discussed in subsequent chapters of this book. However, for you to realize the full potential of the current state maps it is important to keep the following tips in mind (Keyte & Locher, 2004).

Identify Basic Process Boxes First

It is important to identify the basic process boxes before performing the actual walk-through. To do this, your team will need to have discussions about what is the beginning and ending point of the value stream; this discussion is an essential piece of information for completing the VSM. Moreover, it helps to get all your team members on the same page with respect to the larger chunks of activity within the value stream and to identify the internal versus the external clients. These basic process boxes (macro steps) can be modified at any point in the mapping process, if deemed necessary.

Identify Performance Measures

You need to have clarity among your team members with respect to the performance measures your team will collect for each process box. Clarity is achieved by having definitions of the performance measures and specifics about how measurements will be collected and reported. This clarity ensures there is agreement among your team members regarding the measures, and reduces later confusion.

Add Other Information

You should add other information (via visual icons or measures) as you observe the process steps in motion. The important point to remember is that the value stream and process flow maps may not be perfect. Team members need to recognize that maps are an approximation of how a process works and help to identify the issues that prevail. Progress on making improvements will be impeded if the team pursues perfection in these maps.

Do Not Make the Map Unwieldy

You must guard against making the map too unwieldy; start simply and add boxes as necessary. For the VSM where the purpose is to visualize a process at a higher level of abstraction—at the system level—it is important not to include too many process boxes. If the time associated with a process step is relatively short compared with the other process steps, it may be appropriate to combine a group of activities into one process box.

Estimate Performance of Current State

You should estimate the performance of the current state the first time through to get a quick picture of the value stream as it exists. Most service organizations and office or administrative work environments do not have this type of performance data, or what they do have may not be relevant to making lean transformations. To gather precise, accurate data on these processes can take weeks or months. You can save time by getting estimates from the people who perform the tasks or do a short observation to gather estimates. If necessary, you can gather more-precise data later on.

Walk the Value Stream

Your team must walk the value stream to gather the performance data associated with creating value. This tip is important, because it reiterates the need to see, first hand, the implementation of a process. Mapping should not be done only in a conference room. For you to acquire a better understanding of how your process works, you need to walk the process and ask questions, as needed.

Ask Questions

Your team members must ask questions regarding activities and issues they identify to understand potential barriers to designing future states. This is a critical part of the mapping exercise, as what appears to be a solution to a process issue may have negative consequences in other parts of the process. While all the possible consequences may not be known until they are implemented, by asking questions at the mapping stage you may eliminate potential mistakes down the road.

Map as a Team

You should work as a team to map the whole value stream. It is important for all members of your team to see the entire value stream; this objective view is facilitated when your team does the mapping together (with the exception of the team champion or other senior management staff, who will be briefed once the mapping is complete). Moreover, this creates a situation where those not as familiar with a process can ask pertinent questions about why a step in a process is done, whether it adds value, and so forth.

Assign Team Members to Tasks

To keep all members of your team engaged, they should each be assigned to specific tasks, such as team leader, scribe, data recorder, timekeeper, and so forth.

Draw by Hand

Always draw you maps by hand and use sticky notes for process boxes. Creating maps is an iterative process; you may need to modify your first round of drawings as questions are asked and further details are revealed through discussion. Using sticky notes (that can be easily moved around) makes it easier to create the value stream and process flow maps.

MAPPING THE FUTURE STATE

Once the analysis and problem solving steps that are critical to a lean transformation are completed (as described in subsequent chapters), your process improvement team creates a future state map. The steps to create a future state map are the same as you used to create your current state map; in the future state map you have eliminated the waste where possible and entered new performance measures. These new measures are the targets you want to achieve once your improvements have been implemented.

It is important to keep in mind that your first attempt to create a future state map may not be perfect in that you have eliminated all waste, but this should not hamper your effort to create the map. Indeed, it is assumed that improvements will be made continuously as your workers see the results of their process changes and how those changes have affected the performance measures. The goal is to eliminate WAs and continuously improve performance according to the measures identified.

To draw the future state map, your team must identify the specific NVA tasks that can be eliminated. While this may seem straightforward and simple, it may not be. Eliminating some NVA tasks may require that other changes in the process be developed and implemented, some of which may take a longer time to put into place. In our example of applications submitted with a lot of missing or wrong information, it may be determined through analysis that the questions or structure of the form contribute to the problems. This will require changes in the application; those changes may take a while to complete in an organization where layers of bureaucracy must be maneuvered through to complete form revisions. (This is another example of a value stream with too many processing steps.)

Regardless of these types of barriers to quick changes, it is important for your team to identify those steps of a process that can be eliminated immediately without negative consequences, since immediate, positive results provide incentives for workers to continue identifying ways to improve a process. Your future state map may be a *work in progress* for a period as you continuously make improvements. If you reach a point where you are satisfied with your process improvement efforts, then your future state map serves as guideline for standardized instructions on how the process works.

However, before you can begin to map a future state, you need to complete your analysis of the current state to determine root causes of problems and to identify the changes you can implement to eliminate the waste in your process. The next chapter addresses the next step in your lean transformation.

The best way to invent the future you want is to practice imaging it.

— H. B. Gelatt

Appendix

FORM 5.1. Wasteful Activities in [*Name of Organization*]

For each of the WAs identified below, please indicate the extent to which you feel each has an *impact on your organization's (or department's) performance and achievement of outcomes.* Use a scale from 1 to 5, where 1 = has no impact and 5 = has a very significant impact.

Relationship with [*name of organization*]: ❏ Management or Supervisory Staff ❏ Line Staff ❏ Client ❏ Funder
❏ Other:_____

To what extent is your organization's (or department's) performance impacted by these types of wasteful activities?	No Impact	Some Impact	Neutral	Significant Impact	Very Significant Impact	Identify process or process step(s) where the activity has a significant or very significant impact
	1	2	3	4	5	
Waiting						
Convoluted pathways						
Rework						
Information deficits						
Errors or defects in services or materials						
Inefficient work stations						
Extra processing steps						
Stockpiled materials and supplies						
Delivering more services or materials than needed						
Process variation						
Resource depletion						

FORM 5.2. Wasteful Activities and Unacceptable Results in Service Organizations

For each of the WAs and URs identified below, please indicate the extent to which you feel each has an *impact on your organization's (or department's) performance and achievement of outcomes.*

Use a scale from 1 to 5, where, 1 = has no impact and 5 = has a very significant impact.

	No Impact	Some Impact	Neutral	Significant Impact	Very Significant Impact
Waste is found in processes from	**1**	**2**	**3**	**4**	**5**
Waiting					
Convoluted pathways					
Rework					
Information deficits					
Errors or defects in services or materials					
Inefficient work stations					
Extra processing steps					
Stockpiling materials and supplies					
Delivering more services or materials than needed					
Process variation					
Resource depletion					
Unacceptable results of wasteful activities:					
It takes too long to....					
Can't find what's needed.					
Staff have to go get what's needed.					
There are not enough resources to do what needs to be done.					
Clients are dissatisfied.					
Staff or clients can't get answers to their questions.					
Staff are overwhelmed with their workload.					
Staff are frustrated with their work.					
There is a chaotic work environment.					
There is an unsafe work environment.					
Same mistakes are repeated.					
There is a poor quality of services delivered or materials produced.					
There are inconsistencies in delivery of services.					
Staff or clients are "given the run around."					
Staff are not able to get all the work done in allotted time.					
There is too much "red tape" in processes.					
There are handoffs to others (e.g., "It's not my job").					
There is limited clarity in work procedures.					
There are constant interruptions in work.					
There are bottlenecks in workflow.					
Staff duplicate each other's effort.					
There is an unequal distribution of work.					
There is a churn in solving problems (can't get a decision).					
The organization does not produce desired outcomes effectively.					
The work environment is disorganized.					
Staff have to "work around" a defined process.					
There are delays in completion of work or delivery of services.					
Valuable organizational resources are missing.					
Staff reach "dead ends" with no progress made.					
The right hand doesn't know what the left hand is doing.					
Other. Please specify:_____					

FORM 5.3. Process Flow Map Data Collection

Process Mapped: _____

Macro Step: _____

Elapsed Time (CT + WT) = _____

Description of Subtask	Amount of Time*		
	VA	NVA	RNVA
Totals			
Ratios of VA, NVA, and RNVA time to Elapsed time:			

Specify the amount of time allocated to each subtask within the appropriate column, depending on whether the subtask is VA = Value Added; NVA = Non-Value-Added; or RNVA = Required Non-Value-Added.

Problem Solving to Identify Improvement Opportunities

CHAPTER 6 AT A GLANCE

PROBLEM-SOLVING METHODOLOGIES
- Describe the Problem
- Gather Data about the Problem
- Determine the Root Cause(s) of the Problem
  - 5-Whys
  - Cause-and-Effect Fishbone Diagram

GENERATING PROBLEM SOLUTIONS
- Brainstorming
- Benefit vs. Cost/Time Matrix

LEAN TOOLS FOR PROBLEM SOLVING IN SERVICE ORGANIZATIONS
- 5S
  - Sort
  - Set in Order
  - Shine
  - Standardize
  - Sustain
- Workload Balancing
- Visual Controls and Management
  - Visual or Audio Signals
  - Visual Work Instructions

As described in Chapter 5, the value stream and process flow maps of the current state are used to identify WAs and other URs found within the processes. The task of labeling process steps as NVA or required NVA was is the first step in completing an analysis of the process issues. In this chapter, we provide a number of problem-solving approaches that you can use to pinpoint the root cause of a problem and identify potential solutions. In addition, we provide an overview of a number of basic lean tools that you can use as solutions for eliminating WAs and URs in a work environment.

PROBLEM-SOLVING METHODOLOGIES

> *A pessimist sees the difficulty in every opportunity; an optimist sees the opportunity in every difficulty.*
>
> — Winston Churchill

Regardless of the methodology you use to problem-solve, there are a number of steps involved: (1) describe the problem, (2) gather data about the problem, (3) determine the root cause(s) of the problem, (4) generate a solution(s) to the problem, (5) implement the solution(s), (6) verify the effectiveness of the solution(s), and (7) revise and implement new solution(s), if necessary. In this chapter, we will address the first four steps in the problem-solving methodology; Chapter 7 will address the last three steps.

Describe the Problem

Describing the problem is a critical first step. It is important to be specific so the steps you take to identify root cause(s) and solution(s) will result in problem resolution. Some guidelines for writing good problem descriptions include the following (Tapping & Dunn, 2006):

1. The statement should be specific—what the problem is, how big it is, and if it is increasing, decreasing, or remaining the same size.

2. The statement should identify timeframes for its occurrence. Specify when the problem first occurred, how it was identified, and if other events were happening at the same time.

Returning to our concept of WAs and URs first discussed in Chapter 3, WAs and URs are problems that staff, clients, or other stakeholders may experience when they are involved in operational processes of your organization. In the newsletter example mapped in Chapter 5, the UR or problem was primarily related to the flow of work: the process to develop and distribute the newsletter took too long to complete.

Other examples of problem statements, in an abbreviated form, that specify issues with WAs or URs within the different types of service organizations are shown in Table 6.1. As you can see from that table, problems included unrealized collections of child support payments owed to a county government office; a long turnaround time for the preparation, dispensing, and delivery of chemotherapy medication orders in a cancer treatment center; and loss of instructional time that led to underachievement of curriculum goals in a small private middle school.

Gather Data about the Problem

Once your problem statement has been drafted, the next step is to gather as much data about the problem for a more complete understanding of what is happening. Without

TABLE 6.1. Examples of Process Problems from Government, Health, Education, and Social Service Sectors

Sector or Type of Organization	***Government:*** County Government/Office of Child Support Enforcement
WAs or URs Addressed	*Information deficits, extra processing steps, adequacy of resources*: Unrealized child support collections due to unprocessed referrals reduce agency's ability to recover local share dollars expended on behalf of families in receipt of Temporary Assistance for Needy Families (TANF).
Sector or Type of Organization	***Health:*** Cancer treatment centers
WAs or URs Addressed	*Waiting, quality of service, organization of work environment*: Compromised patient safety due to excessive turnaround time for the preparation, dispensing, and delivery of chemotherapy medication orders.
Sector or Type of Organization	***Education:*** Small private middle school in a suburban metropolitan area
WAs or URs Addressed	*Flow of work, quality of service*: Loss of instructional time that led to underachievement of curriculum goals.
Sector or Type of Organization	***Social services:*** Small nonprofit organization in metropolitan area
WAs or URs Addressed	*Information deficits/organization of work environment/adequacy of resources*: Medicaid clients were not signing in/out as required for billing purposes.

Source: The data in row 1 are from "Office of Child Support Enforcement Casebuilding Backlog," by Erie County Government, 2008, Buffalo, NY: Author. Adapted with permission. Retrieved April 11, 2010, from http://www.erie.gov/exec/public/pdf/Office%20of%20Child%20Support%20Enforcement%20Case%20Backlog.pdf

Data in row 2 are from "Doing More with Less: Lean Thinking and Patient Safety in Health Care," by R. A. Porche, Jr. (Ed.), 2006, Oak Brook, IL: Joint Commission Resources. Copyright 2006 by the Joint Commission Resources. Adapted with permission.

Data in row 3 are from "Proven Results Where Results are Most Needed: Case Study 1," by Lean Education Enterprise, Inc. Copyright 2010 by Lean Education Enterprise Inc. Adapted with permission. Retrieved on April 16, 2010, from http://www.leaneducation.com/case-studies.html#study1

Data in row 4 are from "Six Sigma Project: Improve Efficiency of Attendance Verification Pros Program," by Spectrum Human Services, 2009, Orchard Park, NY: Author. Copyright 2010 by Spectrum Human Services. Adapted with permission.

complete information, it will be difficult to identify root causes and potential solutions. In our newsletter example, while some data have been reported in the problem statement (e.g., how many times the newsletter has missed its deadline over the two years and the average number of days it has been delayed), there were other pieces of data that had to be gathered relating to the areas of waste and NVA activity that could be contributing to the UR "It takes too long to create and distribute the newsletter."

In Chapter 5, Table 5.4 offered a useful tool for determining the amount of time spent in NVA activity. As completed, it reveals the percentage of time that is VA, NVA, and RNVA for the *Create* macro step of our newsletter value stream. In this first macro step, you see that a considerable amount of time (91 percent) is NVA, the majority of which is spent waiting for the eleven contributors to submit their article for the newsletter. Based on this information, including the data showing that thirty minutes of the NVA time are spent reminding contributors to submit their article, we see that a considerable amount of time is wasted at the front end of this value stream. In addition to data showing the percentage of time spent in non-valued-added activities (and hence revealing how big an opportunity there is to eliminate waste in a process), there may be other types of data to gather and chart to gain a better understanding as to what is happening and the extent of the problem.

See Example 6.1 for a description of a state welfare agency that provided monetary vouchers to low-income families to subsidize the cost of child care. This example reflects an agency that takes too long to deliver its service (i.e., process applications for subsidy and provide the voucher to eligible parents). Only through gathering additional data regarding the volume of applications (approximately 11,000 per year), the percentage of applications denied (75 percent), and the availability of dollars to pay for the service (sufficient funds to pay for all eligible parents) did we gain a fuller understanding of the problem. Moreover, if these additional data had not been gathered, the solution to the problem would have been different.

At first, you may assume the solution lies in streamlining application processing within the agency through finding where time or other resources are wasted by staff. However, staff spending so much time processing applications that do not meet basic eligibility criteria is a waste of staff time and of no value to the client. The analysis led to a discussion about how parents can better make a predetermination of whether they meet basic eligibility criteria, thus eliminating a large number of applications from parents who do not meet eligibility and eliminating the time staff must spend on these applications. Instead, the staff would be able to spend time on applications from parents, the majority of which would be eligible, thereby resulting in a much shorter processing time to award the parents their childcare vouchers.

Determine the Root Cause(s) of the Problem

Root cause analysis is a method for specifying cause-and-effect relationships. Its purpose is to identify the what, how, and why something happened. Determining the root cause(s) of a problem will lead to recommendations as to probable solutions. Given this, root causes should be identifiable and within the realm of being managed or controlled so the problem will not recur. As a process, root cause analysis involves data collection, cause charting, root cause identification, and recommendation generation and implementation (Rooney & Vanden Heuvel, 2004). You must keep in mind when doing root cause analysis that rarely is there a single causal factor that contributes to a problem. If

EXAMPLE 6.1. Government Welfare Agency Providing Subsidized Child Care

Agency	State Department of Health and Human Services
Service	To provide monetary vouchers to low-income families to subsidize the cost of child care, giving parents affordable arrangements for their children while at work.
Problem Statement	This agency had numerous complaints from applicants that it took several months for their applications to be processed. In some instances, the applicant had to resubmit an application after the original could not be found. While this problem was not the sole unacceptable result of this childcare subsidy program, it represented a significant issue, since it impacted the ability of parents to stay in the workforce.
Current State of Operational Processes and Need for Additional Data	After the team mapped the value stream and determined the percentage of processing time that was VA, NVA, and RNVA, we realized some important pieces of data were missing that could shed additional light on how staff spent their time reviewing applications and making determinations as to eligibility of the parent for the subsidy. These two key questions were asked, and the responses shown: 1. **Question:** Does the state have sufficient funds, which come from federal block grant dollars, to fund all of the applicants for childcare subsidy? **Response:** Yes, for the current number of applicants, the state has the money to provide the subsidies. 2. **Question:** How many applicants for subsidy are processed in a year, and what percentage of applicants meet the basic eligibility criteria? **Response:** There are approximately 11,000 applications submitted in a fiscal year, and approximately 75 percent of the applications *do not meet* basic income guidelines that specify the eligibility of a parent for the subsidy. When this information was provided, the team realized that the case managers spent a majority of their time reviewing applications that could not be approved because the parents exceeded the income level for receiving a subsidy. Furthermore, many of the applications did not include all required income information when first submitted, resulting in case managers having to *work around* this problem to follow up to get the required income documentation. As a result of the way this application process was designed, case managers spent a majority of their time processing applications that would be denied. Therefore, they fell behind in the review process and parents had to wait several months to hear whether they would receive a subsidy. Moreover, case managers were engaged in work that was not rewarding: they had to spend a majority of their time tracking down missing information and informing parents they were not eligible for the subsidy.
Areas of Waste	• Extra processing time • Information deficits • Waiting
Unacceptable Results	• Dissatisfied clients • Frustrated case managers • Unspent federal dollars

> *You miss 100 percent of the shots you never take.*
>
> — Wayne Gretzky

you only address the most visible causal factor, incomplete solutions will result, possibly resulting in problem recurrence.

A number of methodologies have been developed to conduct a root cause analysis. We will present two frequently used methods that are relatively simple and straightforward. The first is **5-Whys** and the second is the **Cause-and-Effect (Fishbone) Diagram**.

5-Whys. This problem-solving analysis tool used in lean transformations was influenced by Taiichi Ohno, who played a significant role in the development of the Toyota Production System. Ohno would direct workers to ask the question "Why?" five times when a mistake or problem occurred in the operation of a process. This questioning empowered workers and enhanced their thinking skills to problem solve and to search for root causes of problems. Ohno believed this was the only way to eliminate the root cause of a problem and thus prevent its recurrence.

The technique is relatively simple: it involves asking a set of *why* questions: Why did such and such happen? Subsequent *why* questions are asked for each response to the previous question; this process eventually gets to the root cause of a situation. It is one of the simplest tools to use in determining root causes and any relationships between root causes of a problem. Once root cause(s) are identified, you have key information to develop a solution(s) to resolve or ameliorate an issue or UR. See Figure 6.1 for an example of 5-Why analysis to determine the root causes for one of the primary URs in our newsletter example: "It takes too long to assemble the newsletter for distribution."

Cause-and-Effect Fishbone Diagram. A fishbone diagram is another type of problem-analysis tool, initially used by Dr. Kaoru Ishikawa, a Japanese quality control statistician; it is often referred to as the Ishikawa diagram. As a tool, it provides a systematic way to look at an effect and the causes that create or contribute to the effect. As its name suggests, its design is similar to that of a fish skeleton. The value of the fishbone diagram is that it offers a way of categorizing the possible causes of problems in an orderly way to search for the root causes. It is particularly useful in a group setting

FIGURE 6.1. 5-Why Analysis

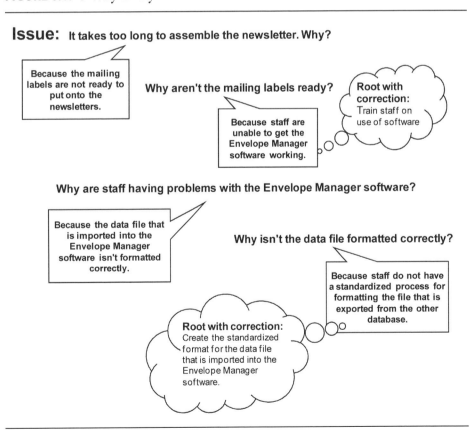

Issue: It takes too long to assemble the newsletter. Why?

Because the mailing labels are not ready to put onto the newsletters.

Why aren't the mailing labels ready?

Because staff are unable to get the Envelope Manager software working.

Root with correction: Train staff on use of software

Why are staff having problems with the Envelope Manager software?

Because the data file that is imported into the Envelope Manager software isn't formatted correctly.

Why isn't the data file formatted correctly?

Because staff do not have a standardized process for formatting the file that is exported from the other database.

Root with correction: Create the standardized format for the data file that is imported into the Envelope Manager software.

where a team brainstorms about causes and there is little quantitative data available for analysis (Simon, 2009). Also, it provides a means to engage in a more thorough exploration of a problem and its causes, which can lead to a more robust solution(s).

Figure 6.2 shows the basic structure of a fishbone diagram and Figure 6.3 provides a specific example of a completed fishbone diagram. To create a fishbone diagram to explain a particular problem or UR within an organization, follow these steps:

1. **State the problem**. The problem is stated in the form of a *why* question; it is the head of the fish or the effect for which causes are to be determined. Stating it as a *why* question will help you brainstorm to find the answer to the question. Before proceeding with the next step, your team must agree on the statement of the problem.

2. **Identify the categories for possible causes**. The rest of the fishbone consists of a line drawn horizontally across a page, attached to the head, along with several other lines (bones) that come out vertically from the main line. Label the major branches with the different categories. The categories used will be specific to your presenting problem and consist of categories that your team agrees are important. However, there are some standard categories that have been used in service work environments, including policies, procedures, people, facility, communication, and system, as shown in Figure 6.2.

3. **Complete the diagram with possible causes**. This is where your team brainstorming occurs to fill in the branches of the diagram. As with 5-Why analysis, a series of *why* questions are asked to identify all of bones on the diagram within each major category branch. This step helps to get to the root causes of a problem.

FIGURE 6.2. Fishbone Diagram Template

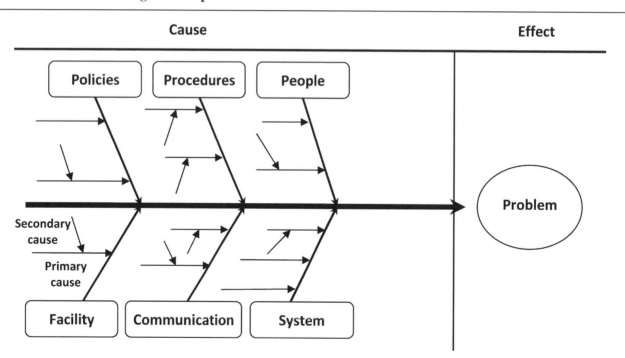

4. **Prioritize the identified causes**. As a team, you should prioritize the causes of the problem, specifying those that are key, since your team will brainstorm to identify solutions for these key causes.

GENERATING PROBLEM SOLUTIONS

Once you gather sufficient data and your root causes have been determined, your process improvement team can engage in further identification of permanent corrective actions to implement. There may be situations where solutions are not obvious or practical, which is why it is important to use a brainstorming process where your team includes key decision makers or their representatives who have the authority to approve the implementation of changes and improvements.

FIGURE 6.3. Fishbone Diagram Example

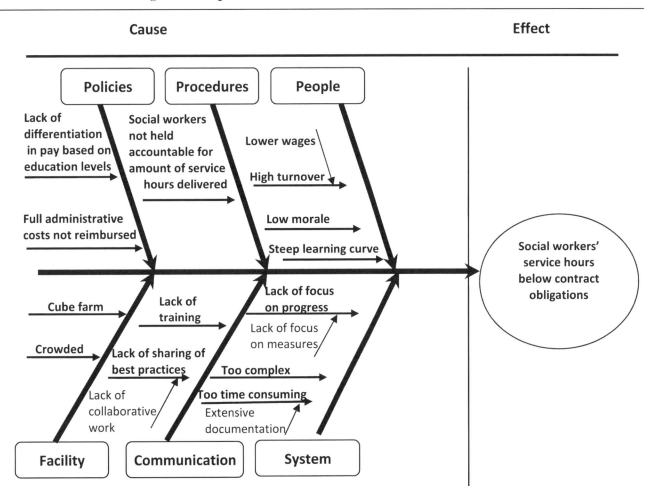

Source: "Buffalo Urban League Six Sigma Project: Family Preservation Services," by F. Siddons, 2010, Buffalo, NY: Buffalo Urban League. Copyright 2010 by Buffalo Urban League. Adapted with permission.

> *We can't solve problems by using the same kind of thinking we used when we created them.*
>
> — Albert Einstein

Brainstorming

Brainstorming is a method used to solicit ideas from members of your team; you should organize these ideas around common themes. As a guide, you can ask these basic questions to identify solutions (Tapping & Dunn, 2006):

1. Is there a better way to design the process?
2. Can the root cause(s) be eliminated?
3. Can negative forces be minimized?
4. Can positive forces be strengthened?
5. Have all possible scenarios been explored?
6. Have others, such as technical experts, customers, clients, and so, on been involved to give their perspectives?

During the brainstorming session, it is important not to criticize any of the ideas, because the purpose is to have as many ideas generated as possible. Subsequent organization and prioritization will result in a refinement of the potential and practical solutions. During this stage, you should identify any of the constraints that might apply to the solutions, such as approvals required, timing of the implementation, cost of the improvements, impact on other processes or people, and so on. It is particularly important to identify any negative or unintended consequences that a change might have on other processes or people, since this knowledge will influence the decisions about solutions to implement. Moreover, a major problem in lean transformations is taking on too many initiatives at once. Therefore, it is critical to have a guide that helps you (1) deselect initiatives down to the ones your organization can really achieve, and (2) align the selected improvement initiatives with company strategic objectives.

> *You can't discover new oceans unless you have the courage to lose sight of the shore.*
>
> — André Gide

Benefit vs. Cost/Time Matrix

A tool you can use to group and prioritize possible solutions is the benefit versus cost/time matrix (Figure 6.4). This matrix groups solutions into four quadrants. Quadrant 1 includes the identified solutions that have the greatest improvement benefit, the associated cost is minimal, and they can be implemented in the short term. In some respects, these solutions might be considered low-hanging fruit and instant winners, in that they represent solutions that are quick and easy to put into place. When you implement these solutions, they provide you with immediate gratification and feelings of success as it relates to making improvements.

The remaining quadrants of this matrix represent additional solutions that vary in terms of their potential for solving your problem and the amount of time and cost associated with implementing your solutions. This is not to say that solutions in these quadrants should not be considered, but some may have to be put on hold if they require a significant investment or time to design and implement. Technology solutions to process issues tend to fall into Quadrant 3 or 4, because technology solutions are more costly and take longer to design, implement, and train workers on their use.

Returning to our example of the childcare subsidy program (Example 6.1), a solution put on the table and placed in Quadrant 3 was to set up kiosk stations in public places throughout the state where parents could answer a few basic questions about income level and size of family to determine if they meet the primary criteria for determining eligibility for the subsidy. The thinking behind this is to have parents do their own prescreening so those who do not meet the basic level of eligibility will not complete and submit applications. With fewer applications to process, case managers will be able to

Figure 6.4. **Benefit versus Cost/Time Matrix**

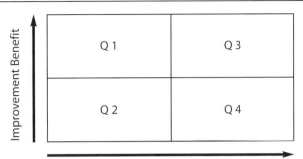

Improvement Cost or Length of Time to Implement

spend their time on reviewing, approving, and setting in place the mechanism for eligible parents to receive the childcare subsidy soon after they submit their applications.

Figure 6.5 shows a benefit versus cost/time matrix for a process improvement within county government. The process involved is the hiring process for filling vacancies within the social services department. There were many low-cost process changes that could be made immediately (e.g., design a new process for dealing with status changes for internal hires, use technology to electronically distribute lists to departments, etc.). However, the improvements that would take the longest time to implement but result in the greatest benefit to eliminate the long cycle time included eliminating the need for the County Fiscal Stability Authority approval, revamping the civil service law, and error-proofing database forms.

Finally, in returning to our example of the development and distribution of a newsletter, we have identified a number of issues and proposed solutions that will have an impact on a number of URs associated with the current process. Table 6.2 shows the issues identified along with the proposed solutions. Together, these solutions, particularly the second (having fewer contributors for each issue of the newsletter) and fifth solutions (standardizing the process for extracting data for address labels) offered

Figure 6.5. **Example of Benefit vs. Cost/Time Matrix to Reduce Cycle Time of Processing Childcare Subsidy Applications**

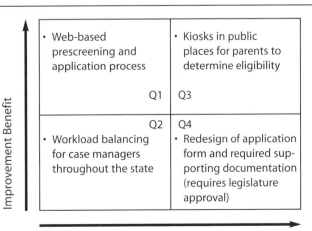

Improvement Cost or Length of Time to Implement

TABLE 6.2. Newsletter Process: Issues and Solutions

Unacceptable Results/Issues	Improvement Ideas/Solutions
(1) Reaching agreement on content within the newsletter and last-minute changes due to events happening at state level that need to be included in a newsletter issue.	(1) Do planning of content only a few weeks before the issue is to go out.
(2) Too many contributors for articles in each newsletter. A lot of time wasted on answering questions and reminding contributors of deadline.	(2) Have fewer contributors for each issue of newsletter. Perhaps rotate who contributes for each issue or only have certain staff always contribute (those that are more efficient in doing this).
(3) Timing with the printer to speed up turnaround time for printing the newsletter.	(3) Once schedule is set, alert printer of the date he or she will get the newsletter to print and expect a one day turnaround time.
(4) Delays in the established schedule for assembly and mailing of newsletter (e.g., waiting for calendar).	(4) Eliminate a printed calendar that needs to be mailed.
(5) Getting labels printed from Envelope Manager software.	(5) Standardize the process for extracting data from state database for the addresses and putting it into the right form so that labels can be produced from Envelope Manager software.
(6) Change in USPS requirements for bulk mailing.	(6) Maintain close contact with USPS to ensure that updates in system are known.

easy-to-implement solutions to problems that consistently slowed down the process to create and distribute the newsletter.

LEAN TOOLS FOR PROBLEM SOLVING IN SERVICE ORGANIZATIONS

In keeping with the lean philosophy, the following provides you with a number of tools to solve operational problems or issues affecting your organization's ability to meet clients' needs in the most efficient and effective manner. These tools include **5S**, **Workload Balancing**, and **Visual Controls**. While these are not all of the lean tools you can use to eliminate waste in your work environment, they represent tools relevant to service organizations that are easy to implement. Furthermore, there are components of each of these lean tools that overlap each other, which will become apparent in our discussion of them.

5S

Implementing 5S within work areas is sine qua non within an organization that is focused on streamlining its processes and improving operational efficiency. The essence of 5S is to add visual controls, standardization, and organization to your organization's operational processes (Hirano, 1993, 1995; Osada, 1991). In your organization where you have repetitive tasks, 5S provides an approach to streamline these activities and eliminate simple waste. It provides a method to systematically keep work areas clean and organized while ensuring safety and providing a foundation for the implementation of other lean solutions within your organization. Furthermore, it is a lean tool that is easy to implement. Everyone in a work environment can get involved in the implementation of 5S.

What is the 5S improvement tool? Simply put, it involves a number of systematic steps to organize your work environment: (1) sort, (2) set in order, (3) shine, (4) standardize, and (5) sustain. The simplest way to describe the application of this tool is with respect to a workstation. No doubt we have all seen workstations cluttered with paper, food, files, personal items, and so on. A question we often ask is, "How can Employee A get any work done in there?" Just think of the time wasted trying to find something needed to complete work; the build-up of dust and perhaps insects as a result of food left out, creating health hazards; and the clutter on the floor that can create trip hazards. None of these situations fosters a good work environment. That is where 5S comes into play.

Sort. Your first task in approaching such a workstation is the first S, to sort: go through all items and identify them as (1) items never used (red tag), (2) items used infrequently or that have questionable use (yellow tag), and (3) critical items used frequently (green tag). After discarding those items that are obviously trash, the next step is to determine the disposition of the tagged items. This can include returning to the area, disposing of them, or donating them to charity.

Set in Order. After your sorting process is completed, the second S is to put the essential items in a place, or to set them in order so they are best located for the work at hand, with the most frequently used items in a close location. This requires that you label or visually mark off the areas where items belong, whether in file drawers or on spaces on the floor. With the visual marking of places where items belong, you will easily see if items are missing or out-of-place.

Shine. The third S is for shine: clean up the work area and all items that are placed there. Shine can also mean to remove or diminish the sources of disorganization within a work environment. Once basic cleaning is done, it must be kept up on a regular basis. This ranges from having a designated day for spring cleaning when all your staff participate to having a schedule of cleaning completed by others for areas or items within the work environment such as carpets and floors, windows, outside grounds, common areas, rest rooms, and so forth. You should have a cleaning plan that establishes what is to be cleaned, how often, and by whom. Individual employees also must commit to keeping their own areas set in order, clean, and safe.

Standardize. Once your work area is shinning and set it order, the fourth S, standardize, is already well under way—everything in its place and a place for everything. You must properly label items (using color coding for them) so that it is easy for outsiders or new employees to locate them and to know where they belong. Another component of standardization is with regard to work processes—having a standard way to carry out a process from start to finish. This means that you clearly specify the work content, timing, sequence, and outcomes of work tasks. This reduces variation in the way the same work is completed, either by the same person over time or across several people doing the same work. Knowledge of the standards for a work process will also make it obvious when standards are not met. With no ambiguity about who provides what to whom and when, automatic quality checks of the service or product as it flows through its process steps are built in. You will be able to identify errors or mistakes earlier in the process so they can be corrected before you move to the next process step. Visual work instructions provide a valuable lean tool in creating standardized work processes that can be easily followed—even by new employees. The importance of visual controls and management of work processes will be further discussed later in this chapter.

Sustain. Finally, your organized work area and standardized work processes must be sustained over time, and that is the fifth S. Sustaining this system of visual controls, standardization, and organization is probably the most difficult and requires a system of education and monitoring, at least until these expectations become embedded within your work culture and the activities required to sustain the organized work environment become normative. Therefore, your organization must establish a system of auditing work areas to ensure that standards are maintained. An example of a 5S audit is shown in Figure 6.6.

Figure 6.6. Form for a 5S Work Area Audit Checklist

Work Area: _____ Assessor: _____ Date: _____

Objective: To determine the status of a work area in meeting our company standards of a clean, organized, and safe work environment.

Step	Assessment	MS	NI	NF	NA
SORT	No infrequently or unused materials or paperwork in area.	—	—	—	—
	No outdated notices or project-related information posted.	—	—	—	—
	No excessive personal items in work areas (20 is a guideline).	—	—	—	—
	No trash and disorderliness or untidiness.	—	—	—	—
SET IN ORDER	Key items needed to perform work can be located without having to search.	—	—	—	—
	Storage locations or files (hard copy and electronic) are labeled for quick retrieval of information.	—	—	—	—
	Work-in-process is in clearly marked trays or files.	—	—	—	—
	Materials or files are stored in designated areas.	—	—	—	—
	Work area is clear of obstructions and trip hazards.	—	—	—	—
	Fire extinguishers are clearly marked and accessible with no obstructions (three-foot clearance).	—	—	—	—
	Supplies and tools are organized, stored, and labeled properly.	—	—	—	—
SHINE	Workstations are clean and orderly at the end of the day.	—	—	—	—
	Floors are free of paper piles, swept clean, and clear of trash.	—	—	—	—
	Trash and recycle bins are accessible and not overflowing.	—	—	—	—
	Computer keyboards and monitors, shelves, ledges, desks, fans, etc. are free of dust and debris.	—	—	—	—
	Cables are neatly tied and not left dangling or exposed to create trip hazards.	—	—	—	—
STANDARDIZE	Color codes exist and staff understand them.	—	—	—	—
	Documentation and information is formatted as stated in each program's policy and procedures manual.	—	—	—	—
	There are not a large number of different files or documents for each work operation.	—	—	—	—
	A minimal number of computer screens must be accessed to process regular work.	—	—	—	—
	A minimal number of forms or files must be accessed to process a request.	—	—	—	—
	Common language or common terms are defined.	—	—	—	—
SUSTAIN	Regular workplace meetings are conducted to allow staff to discuss improvement ideas.	—	—	—	—
	Suggestions and improvement ideas are reevaluated regularly.	—	—	—	—
	Supervisors review work practices with staff.	—	—	—	—
	Staff routinely conduct workstation assessment and perform clean-up as necessary.	—	—	—	—
	Clean-up responsibilities and frequencies are communicated and posted.	—	—	—	—
	Sufficient clean-up time has been allotted.	—	—	—	—

MS = Meets Standard, NI = Needs Improvement, NF = No Foundation in Place, NA = Not applicable.

©2008 KeyStone Research Corporation

While this approach to your organization's work environment may seem rigid and extreme, there is a sound rationale to its use. Implementing 5S, putting an entire work environment in order, and sustaining it over time

1. improves safety and ergonomics for your workers;
2. introduces your staff to a lean tool that is simple to understand and implement;
3. lays the foundation for a lean work environment;
4. reduces your costs by eliminating wasted time, wasted space, and wasted effort;
5. facilitates the smooth flow of work and increases productivity;
6. reduces your employees' stress and improves their morale;
7. creates a sharper workplace appearance and impresses your clients;
8. utilizes your space better and reduces a cramped, cluttered, workspace;
9. provides a systematic process for improving your work environment; and
10. shows the importance of standardization in creating an efficient and effective work place.

Workload Balancing

In lean manufacturing, this concept is referred to as line balancing because it provides a method for engineers to set the pace of work flowing through an assembly line to meet a customer demand. The goal is to determine the amount of time to be allocated to the work content (called "takt" time in lean terminology, from the German word *taktzeit*, or cycle time) so it perfectly matches the amount of time spent meeting your clients' needs. In other words, if your organization operates 480 minutes a day and your clients submit two hundred applications for a service that must be processed in one day, then the takt time is two minutes; that time should be allocated to the processing each application to meet the demand.

In service organizations, however, the notion that your work can be treated in the same way as assembly lines producing products does not engender a positive response. Visions of time and motion studies and a regimented work environment, where your staff are strictly controlled with respect to how they spend their time, does not resonate with staff who know their work content is subject to more unknowns. Moreover, for some services delivered to clients, the more time spent with them (rather than a strictly controlled amount of time, regardless of situation), increases the satisfaction. As an example, if a caseworker in a hospital has a responsibility to do aftercare planning for frail patients that leave the hospital, the amount of time it may take to address all the questions and issues associated with this aftercare planning will vary greatly, depending on the patient's presenting condition and sources of community support. Therefore, if a case manager is restricted in the amount of time he or she can spend on a patient's aftercare plan, lower quality in the plan and reduced patient satisfaction could result.

Regardless, there are some important concepts associated with work balancing. Takt time can be used within your service organizations to determine a measurable unit of work that can be completed is a specified amount of time, the amount of staff hours that will be needed to complete the work within a specified timeframe, and a means to schedule work so that your organization will stay on track to meet your performance goals.

To determine these amounts of time and staffing requirements, your service organization will need to gather data about how long it takes to complete different units of work. This is where value stream mapping of the current state comes into play. These maps show your cycle times, wait times, and total elapsed time for a work process. It also shows the percentage of time that is NVA. With this information in hand, you can establish a goal with respect to a takt time equivalent. Again, taking our example of the case manager in a hospital, if it is determined that a case manger takes an average of two hours to complete an aftercare plan and in any single day the case manager is likely to work with six cases for an hour each (i.e., the case manager's effective work time for a day is six hours), then the turnaround time for the aftercare plan is two days. In the span of a month (twenty working days), the case manager will be able to complete sixty aftercare plans (i.e., 120 working hours divided by two hours per aftercare plan = sixty plans per month). If the hospital averages 360 patients each month that need aftercare plans, then to meet the demand and have a two-day turnaround time, the hospital should have six case managers.

Another way that your service organization can use these concepts is in the allocation of units of work across several staff, so that the flow of work remains steady and bottlenecks do not occur. When bottlenecks occur, results may be wait times for work to move to the next step of a process, down time for staff who are not getting their work from upstream, and increase of the overall elapsed time to complete the subprocesses across a value stream. As an example of this situation, take an organization that produces a written report on a quarterly basis. Each report may take an average of forty-five hours to complete, and there are three staff persons that contribute to the finalization of this report. If the turnaround time from the start of the report writing to when it must be out the door is one week (i.e., five working days), then to ensure the report is completed on time, for a perfectly balanced distribution of work, each staff person should spend fifteen hours on his or her portion of the work; this averages three hours per day for the five days. To accomplish this, the work of writing the report should be chunked into fifteen-hour units that can be assigned to each staff person.

However, in the real world this is not always easy to do, particularly if you have different skill levels required for the work that must be completed. Or you may have a situation where one of the three staff persons is absent and cannot complete the assigned portion of the work. This is where the notion of cross-training and flexing labor comes in as a way to continue the flow of work with fewer individuals. In our example of the report that must be written by two staff persons instead of three, this means that the two staff persons must now pick up the remaining work tasks and divide it between them if they are to meet the deadline of one week to complete the report. They can only do this if they have the skills to complete this new work. If the staff do not have the skills, then the UR will be a delay in completing the report since the labor could not be flexed to meet the deadline.

In essence, as a service organization applying these lean concepts, you will be able to determine the capacity of your current workforce, the reasonable work expectations with respect to how much work can be completed over a specified time span for each worker, and the amount of labor (and cost associated with this labor) that you will need to meet your client demands. It is not uncommon for service organizations to have clients who complain that it takes too long to receive a service or to have staff who are overwhelmed and frustrated because they do not seem to have enough hours in the day to complete all the work that is required of them; both of these are URs of the way work processes are designed and implemented. However, this type of systematic way of viewing work content reinforces the importance of organizational time and how it should be spent on VA rather than NVA activities. Furthermore, it enables you to convey the rationale for establishing expectations for worker performance. Also, you can use these data about organizational capacity to justify an under-resourced situation, maintain your current staffing levels, or take the position that your organization can do more with less.

Visual Controls and Management

Workplace visual controls are a way to manage work activity; they require you to use signs, labels, and color-coded markings so that in a matter of minutes anyone unfamiliar with a process can know what is going on, understand the process, know what is being done correctly, and what is out of place. In essence, the status of the system can be understood by everyone at a glance.

There are two types of visual controls that are relevant to service organizations:

1. Visual or audio signals that display information about operation of a process
2. Visual work instructions that show the steps of a work process

Visual or Audio Signals. Your use of visual or audio signals helps you keep work running as efficiently as designed. These signals help to alert your staff that a work activity needs to be completed, or signal that something is not right about a process, indicating it needs to be addressed. With the use of visual or audio signals, the time it takes to address either of these issues will be reduced, since the information needed to trigger an activity is not hidden and is out of sight.

For example, your organization probably has some type of process to replenish or store supplies and equipment used by staff. In a lean work environment, visual signals are used to show not only the placement of these supplies and equipment, but also the points at which particular items need to be replaced and the order amounts. If we take an office supply room, there would be different areas clearly marked as to where the different color and sizes of copy paper are stocked. In addition, there would be pull cards placed in the stacks of paper that signal the point at which the paper needs to be ordered, from whom it is ordered, and the amount to be ordered. The determination of amount is based on data that provides an estimate of how much paper is used over a designated period. If your organization has a cycle of ordering supplies monthly, then you must have an estimate of the amount of each supply used in a month. In that case, you must always have a minimum of one month's supply on hand, and your reorder card must be placed at the point where you reorder to meet the requirement that there always be a one month's supply on hand. Thus, if there is a one-day turnaround time for delivery of paper supplies, you place the pull card at the point where there is only one month's supply left.

Another example of the use of visual controls in a work environment is with the placement and use of shared equipment. If your organization has an LCD projector and laptop computer that are used by multiple staff, it is important to have a set of visual controls in place to show when the equipment is going to be used, by whom, and for how long. A sign-up sheet on the cabinet where the equipment is kept is a way of doing this.

While the use of visual controls as discussed above may seem trivial, it is important to remember the role they play in reducing the amount of time wasted, whether it involves looking for misplaced or lost items or waiting to complete a work activity because the needed supplies or equipment are not available. Also, when visual signals are used to identify an item of critical importance (e.g., a red flag is used on an email to alert the receiver of its importance and the need to open, read, and respond to the email ASAP), then the response time will be reduced or problems addressed before they get out of hand.

Visual Work Instructions. Visual work instructions provide you with the means to clearly display process steps or work activities so there is consistency in the way work is done, whether by different people or the same person over time. They also provide a way to ensure that your service or product meets established standards. While we are all familiar with work instructions in procedural manuals, often these manuals are tucked away in bookshelves and file cabinets, making them less likely to be used. As a result, when work instructions are not readily available and easily understood, it does not take long for your staff to establish their own way of doing their work. If there is little oversight of the work completed and few quality checks, the likelihood of process variation, quality errors, and dissatisfied clients are increased.

Example 6.2 describes a service organization that was responsible for warehousing, packaging, and distributing health and safety educational materials to a group of trainers who used these materials in their educational seminars. It is evident in this example that serious problems existed with respect to the content of these packaged materials, resulting in trainers being dissatisfied when they would receive these packages that had missing items or outdated material. The photos in this example (see figures 6.7a and 6.7b) provide a good example of how work instructions can be displayed to ensure the work required to assemble the packaged materials is clearly shown in a step-by-step sequence. Furthermore, this made it easier for the organization to hire temporary workers to assemble these materials during peak times, since the work tasks were clear, reducing the amount of time it would take to train someone on the assembly process.

Example 6.2. Warehousing, Packaging, and Distributing Health and Safety Materials

Agency	ABC Organization, a service firm that provides warehousing, packaging, and distributing educational materials.
Problem Statement	• To establish a system for ordering, warehousing, packaging, and distributing a health and safety kit to childcare providers throughout the state.
Current State of Operational Processes	• The responsibility for distributing health and safety kits to childcare providers throughout the state was transferred from one vendor to ABC.
	• The previous vendor passed information to ABC regarding the content of each of the health and safety kits, and also sent existing kits (prepackaged) to ABC.
	• Some of the kits received had incorrect or missing items, unbeknownst to the ABC staff. Therefore, the ABC staff put together new kits, based on the contents of the ones they received, resulting in kits that were shipped with incorrect or missing items.
	• Childcare providers received kits that did not have some of the items or had items that were not to be part of the kit. They were dissatisfied with the services received from ABC because of these errors.
	• The ABC team received conflicting information from the statewide agency that was responsible for developing the content for these kits with respect to the specific items that were to be included in the separate kits.
	• The ABC team was frustrated because staff did not know the steps to take to correct the problem.
Areas of Waste	• Errors or mistakes: the kits were not put together correctly.
	• Rework when missing items had to be shipped to childcare providers.
Problem Solution	• Determination as to the right contents of the kits.
	• Creation of visual work instructions to ensure that the kits were put together properly and packaged in the correct way (see Figure 6.7).

FIGURE 6.7a. Visual Instructions for Final Assembly of Health and Saefty Kit

FIGURE 6.7b. Visual Instructions for Assembly of Subpart of Health and Saefty Kit

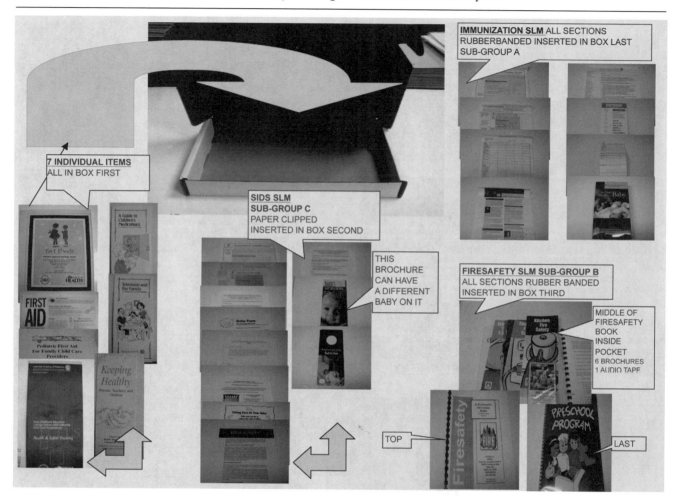

CHAPTER 7

Implementing Improvements and Tracking Results

CHAPTER 7 AT A GLANCE

INITIATING YOUR CYCLE OF CONTINUOUS IMPROVEMENT
- Kaizen Event
- Action Planning Tool

DISPLAYING RESULTS OF YOUR IMPROVEMENT EFFORTS
- A3 Report

A common way of organizing problem-solving activities and implementing improvements was depicted in Chapter 2, Figure 2.2. This figure shows the continuous change cycle in terms of four major phases:

1. Phase I: Planning for a change
2. Phase II: Implementing the planned changes
3. Phase III: Evaluating or assessing the results of the implementation
4. Phase IV: Taking action based on learning that occurred in the evaluating or assessing phase

As previously discussed, this cycle of change and continuous improvement is based on the scientific method for experimentation to test hypotheses. Our earlier chapters laid the foundation for stating the problem, gathering an initial set of data to empirically describe the current state, and hypothesizing as to the root causes of the problem or process issue. With this knowledge, your next step involves the development of an intervention (i.e., change in the current state of a process), to resolve the problem or process issue.

INITIATING YOUR CYCLE OF CONTINUOUS IMPROVEMENT

The following describes an approach to initiating improvement efforts within your organization and a systematic way to plan for changes that will be implemented.

> *If you think you can, you can. And if you think you can't, you're right.*
>
> — Mary Kay Ash

Kaizen Event

In lean transformations, it is common to hold a **kaizen** event—a concentrated improvement effort within an organization. Kaizen is a Japanese term used to describe the chain of events associated with improvement efforts. "Kai" means to take apart and "zen" means to make good. Kaizen has become synonymous with continuous improvement (Tapping & Dunn, 2006). Typically, it is a group activity focused on a specific value stream or process that lasts five days, in which your team

1. receives training in lean concepts and methods, if necessary;
2. creates the value stream and process flow maps;

3. gathers accurate data on cycle times, number of people involved, and resource costs;

4. identifies and prioritizes the opportunities for improving the process in one or more ways, based on a root cause analysis and the desired future state map of the process;

5. develops the initial action plan to implement the improvement; and

6. tracks the results over time.

Following the implementation, your team will spend additional time tracking results; reporting on the results to management and other organizational staff, as appropriate for the particular improvement effort; and continuing to make improvements based on the assessment of the results of the improvement effort. The amount of time spent on the initial planning for a kaizen event and the follow-up will vary depending on the focus of the improvement efforts. Moreover, who is involved in each of the phases of an improvement effort will vary. Generally, if an overall value stream is the focus, then it is likely that management personnel are involved. Once specific issues within your value stream are identified, then a kaizen event for work teams and team leaders is established (Marchwinski et al., 2008).

The first four steps that a team goes through in a kaizen event have been described in earlier chapters of this book. In this chapter, we will focus on the tools that can be used to (1) prepare an action plan that details the implementation of improvement efforts, (2) track and graph results of improvement efforts, and (3) report results to team members, management, and other external stakeholders.

Action Planning Tool

The plan that you develop for your lean transformation must specify the **what**, **how**, **who**, and **when** of your change efforts. In addition, you must maintain status notes with regular updates as the progress made on your plan changes. These status notes also can specify any modifications of the action plan that are made. See Form 7.1 for the **Action Planning Tool** template that can be used to document these details of the improvement effort.

This form provides a systematic way of detailing the change effort and ensuring that all involved are aware of their responsibilities and timeline for meeting their commitments. Appendix 7B at the end of this chapter provides an example of a completed Action Planning Tool for a lean transformation case involving the state childcare subsidy program that provided vouchers to assist working parent(s) in the payment of childcare costs. This tool has a number of components:

1. **Name of program or service.** This is the title of the specific program or service delivered by the organization. Since service organizations may operate a number of programs, this title identifies that part of the organization that will be the focus of the process improvement effort(s). This program may be a value stream in and of itself, or there may be multiple value streams and processes within the identified program or service. In the example in Appendix 7B, the name is Subsidy Program.

2. **Date.** This is the date on which the action plan was developed. If the planning occurs over several days or more, then a date range can be put in this section of the Action Planning Tool.

> *Opportunity is just there, to see or not, to use or pass by. We have to do the hard work.*
>
> — Bonnie Neugebauer

FORM 7.1. Action Planning Tool Template

Name of program or service:			Date:	

Process mapped:

Goal of process improvement:

Areas of waste or UR addressed:

WHAT: Objective*	HOW: Action Steps*	WHO:		WHEN: Timeline
		Lead	Other Stakeholders or Partners	
1.	1-a			
	1-b			
	1-c			
STATUS NOTES				
Date	Comments			

*Add additional objectives, action steps, and pages, as necessary.

3. **Process mapped**. For the purposes of this action plan, the improvements to be implemented should be detailed for each subprocess within a value stream that is targeted for a kaizen event; this represents a set of activities that are manageable. In our example in Appendix 7B, there were a number of teams that focused on improvements for specific processes. Our example included only one part of the Action Planning Tool that focused on this process: Completion and return of the Childcare Arrangement form.

4. **Goal of process improvement**. The information in this section of the Action Planning Tool provides the overarching goal, or the desired result of the improvement effort. In our example in Appendix 7B, the goal associated with the first process identified above was specified as "To ensure the accurate completion and proper return of the Childcare Arrangement form to the designated case worker."

5. **Areas of waste or URs addressed**. This section of the Action Planning Tool identifies the specific problems or issues that the team identified during its initial mapping and analysis efforts. Again, once problems are identified and root cause analysis is completed, a team will be able to brainstorm for solutions to implement. In the Appendix 7B example, the area of waste identified was "Required rework on the Childcare Arrangement form as a result of parents having incorrect authorization worksheet, which they use to complete the forms that are submitted as part of the subsidy application."

6. **WHAT: Objective**. The specification of the objective(s) provides another level of detail regarding the focus of the process improvement effort and the result

that is expected. There may be multiple objectives stated; how many are detailed in an action plan will depend on the size and scope of the improvement effort. In our example in Appendix 7B, there was one objective stated: "To ensure that the Childcare Arrangement form is completed accurately by provider, given to parent to submit to assigned case worker for accurate authorization assignment."

7. **HOW: Action Steps**. In this section of the Action Planning Tool, the specific steps that must be taken to accomplish the stated objective(s) are outlined. These are more concrete with respect to actions that, once completed, will result in the achievement of the objective associated with the action steps. In our example in Appendix 7B, the four action steps included these:

   a. Develop a checklist and update the Childcare Care Arrangement form to make it easier to complete accurately.

   b. Inform childcare providers and parents of the new process for completing the Childcare Arrangement form.

   c. Distribute the Childcare Arrangement forms to childcare providers.

   d. Update information or checklist on the Childcare Arrangement form.

8. **WHO: Lead and Other Stakeholders or Partners**. In this section of the Action Planning Tool, the specific persons responsible for completing the action steps are identified. In our example in Appendix 7B, the names of these individuals were provided. If there is a situation where persons are identified to complete action steps, but they are not part of the improvement team doing the planning, then the leader of the team, or another designated team member, must be assigned the responsibility of communicating this responsibility to the person.

9. **WHEN: Timeline**. This section of the Action Planning Tool specifies the due date for the completion of each action step. It is important to be realistic about timelines and to modify them as necessary when status updates are done. Ideally, timelines are realistic and individuals are able to keep their commitments; there may be unexpected occurrences, however, that impact the ability of the responsible team members to adhere to the timelines. What is important in the Action Planning Tool is that the timelines are modified as appropriate throughout the implementation process.

10. **STATUS NOTES: Date and Comments**. This is the section of the Action Planning Tool that details the discussions that occur among team members as they touch base with one another on a regular basis during the implementation of the improvements. This section is open-ended and can be used to capture the most important pieces of information that reveal the status of the implementation, whether positive or negative with respect to the action steps taken. There may be information provided during these status updates that result in modifications of the action plan. It is important to document any roadblocks or challenges faced by the team so that further strategizing can take place and the action plan can be changed accordingly. For our example in Appendix 7B, the status notes specify the additional suggestions made for modifying the Childcare Arrangement form and noted there was a formal process within the agency that had to take place before the new form could be promulgated throughout the system.

DISPLAYING RESULTS OF YOUR IMPROVEMENT EFFORTS

As process improvements are made, a primary goal of any lean transformation is to reduce significantly the percentage of NVA time that it takes to complete a process, in comparison to the percentage of VA and required NVA time. Therefore, one method of tracking results is to provide a comparison, over time, as to the percentage of elapsed time that falls into each of the three categories. As depicted in Figure 7.1, if the improvements have the intended results, the line graph shows a decline in the percentage of time that is NVA and an increase in the percentage of time that is VA; the required NVA time remains fairly steady. However, if possible, steps can be taken to reduce the required NVA time. While these changes over time are displayed with a line graph, a bar chart can be used instead.

Furthermore, over time the specific performance measures for a particular process improvement effort can be tracked to determine the results from the expected improvements. Returning to our example of the newsletter process, Figure 7.2 shows the

FIGURE 7.1. Tracking VA, RNVA, and NVA Time

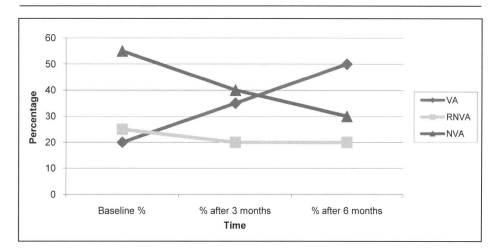

FIGURE 7.2. Newsletter Performance Measures Over Time

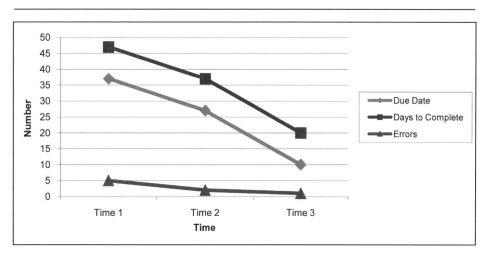

expected changes in the performance measures, as detailed in Chapter 5, Table 5.2, where these performance measures were identified:

1. The newsletter is **mailed on the specified due date**, where performance is measured by the number of days before or after (+ or –) the specified date for mailing.
2. The newsletter preparation process takes **ten working days to complete**, where performance is measured by the number of days (+ or –) needed to produce the newsletter compared to the expected number of days.
3. The **newsletter has no errors** in it, where performance is measured by the number of errors within the newsletter.

If the results were achieved as expected, then the newsletter would be mailed on the due date, rather than being chronically late (i.e., the line graph would show a decline on this measure); the number of days to complete the newsletter would decline, going from the current state of forty-seven days to ten days, and the number of errors would decline over time, with the goal being zero errors.

Other formats for displaying results of improvement efforts are shown in Table 7.1 and Table 7.2. The first table provides a tool that can be used to report **Success Stories**. This list summarizes the key information about a process improvement effort:

1. The program name.
2. The action steps completed.
3. The specific areas of waste or URs addressed.
4. The date the improvement was implemented.
5. The performance measures associated with each process improvement effort, specified in terms of amount of improvement (e.g., reduced time by three hours, or saved $50,000 in operational costs).

TABLE 7.1. Success Stories Template

Process Improvements: ABC Organization

Action Step #	Action Step Completed	Areas of Waste or URs Addressed* ✓				Date Implemented	Savings/Performance Measures		
		WA/UR 1	WA/UR 2	WA/UR 3	WA/UR 4		Time (Hours)	Money	Other

* Specify the areas of waste and/or unacceptable results addressed in your process improvement efforts, e.g., waiting, rework, information deficits, it takes too long, client dissatisfaction, etc.

Table 7.2 shows a format comparing the **performance measures at the current state versus performance measures at the future state**; this table is another way to display the results of improvement efforts. In this table, the following information is captured:

1. The name of the process you focused on in the improvement effort.
2. The specific performance measures (metrics) you used for the improvement effort.
3. The value of the performance measure when you gathered data for the current state.
4. The value of the performance measure when you gathered data for the future state, after you implemented the improvements.
5. The amount of improvement, as a percent of change from your current state to the future state.

While there are any number of ways to display the results of your improvement efforts, the key to continual improvement is to collect before and after data and provide some type of summary as to the results. If this part of the improvement cycle is not completed, your process improvement team will not receive the pat on the back that is an

TABLE 7.2. **Comparing Current State versus Future State Performance Measures for Application of Lean Health-care Practices in Four Hospital Work Groups**

1. Hospital billing process from receipt of voucher to transmitting claim or posting payment

Metric	Current State	Future State	Improvement
Number of process steps	16	9	42%
Process time (minutes)	86	37	57%
Lead time	5 days	2 hours	90%

2. Endoscopy procedure process from patient arrival to discharge

Metric	Current State	Future State	Improvement
Process time (minutes)	178	131	26%
Patient wait time (minutes)	81	11	86%
Lead time (minutes)	260	142	45%

3. Office visit process from patient arrival to completed report

Metric	Current State	Future State	Improvement
Number of process steps	24	6	75%
Process time (minutes)	179	58	68%
Lead time (days)	34	1	97%

4. Scheduling process from physician inquiry to scheduled appointment

Metric	Current State	Future State	Improvement
Number of process steps	14	8	42%
Process time (minutes)	69	18	70%
Lead time (days)	34	3	90%
Percent rework (rescheduling) appointments	25%	2%	92%

Source: From Administrative Lean™ of Lean Concepts, LLC. Retrieved from http://leanconcepts.com/case_studies.htm. Reprinted with permission.

important part of maintaining momentum and enthusiasm for implementing other improvements. Also, if your improvement efforts are not producing the results as expected, then your summary of results provides the motivation to reassess the condition, its root causes, and other possible improvements.

Additional examples of process improvement efforts and the results from them are shown in Table 7.3. This table displays a continuation of the examples of problem statements from service organizations in government, education, health care, and social services, as presented in Chapter 6, Table 6.1. The government Office of Child Support Enforcement reduced their backlog of unprocessed referrals from 7,281 to 2,635, a 64 percent improvement. In the cancer treatment center, the improvements included a reduction in turnaround time for chemotherapy treatments from sixty-one minutes to forty-seven minutes, a decrease in the number of process steps from thirty-two to sixteen, an increase of 30 percent in the number of chemotherapies, and a 50 percent increase in the number of face-to-face contacts with patients. The small private middle school recovered 120 hours of instructional time per teacher over the nine-month school year, had higher levels of staff involvement in planning and scheduling of school and team level activities, increased the student's exposure to the curriculum, and improved learning at the student level. The social service agency reduced the amount of time spent recovering missing data for Medicaid billing by 1,734 hours, a savings of $54,000.

It's not whether you get knocked down. It's whether you get up.

— Vince Lombardi

A3 Report

While the tables and graphs used to track the results of improvement efforts are one element of a report, they do not tell the whole story as to the focus of your effort, implementation plan, and results. To succinctly organize this type of information to tell the whole story, the **A3** report has been used in lean transformations for both managing the entire process and for reporting on the results and next steps of improvement efforts (Shook, 2008).

A3 derives its name from the size of an international piece of paper of 11 by 17. In lean transformations, an issue or problem and efforts to resolve it in a succinct way that is easily grasped is captured on a single piece of paper. The A3 report enables all that are touched by the issue to see it through the same lens.

There are a number of key elements of an A3 report. Its format and exact content may vary, however, depending on an organization's preferences and the problem or issue being addressed. Form 7.2 shows a template that can be used for an A3 report. The key elements of an A3 report generally include the following (Shook, 2008, p. 7):

Title: Name of the issue, problem, or theme of the effort.

Owner & Date: Identifies the person, department or agency who owns the problem or issue and the date of the latest revision of the A3 form.

1. **Background**: Establishes the business context and importance of the issue.
2. **Current Condition**: Describes what is currently known about the problem or issue.
3. **Goals or Targets**: Identifies the desired outcome(s).
4. **Cause Analysis**: Analyzes the situation and the underlying root causes that have created the gap between the current situation and the desired outcome(s).
5. **Proposed Countermeasures**: Proposes some corrective actions or countermeasures to address the problem, close the gaps, or reach a goal.

Table 7.3. Examples of Process Improvement Projects from Government, Health, Education, and Social Service Sectors

Sector or Type of Organization	*Government:* County Government, Office of Child Support Enforcement
WAs or URs Addressed	*Information deficits, extra processing steps, adequacy of resources:* Unrealized child support collections due to unprocessed referrals reduce agency's ability to recover local share dollars expended on behalf of families in receipt of Temporary Assistance for Needy Families (TANF).
Root Causes and Solutions	Process improvement tools were utilized to address the issue of unprocessed referrals. The following causes were addressed: low staffing level, low quality of referrals, and manual processing of information.
Results	Reduction of backlog of unprocessed referrals from 7,281 to 2,635, a 64 percent improvement.
Sector or Type of Organization	*Health:* Cancer treatment centers
WAs or URs Addressed	*Waiting, quality of service, organization of work environment:* Compromised patient safety due to excessive turnaround time for the preparation, dispensing, and delivery of chemotherapy medication orders.
Root Causes and Solutions	Several lean projects were initiated to improve the turnaround time, including (1) a daily lean management system that provides visual feedback about goals and current performance levels, (2) a 5S and primary visual display that communicates information about quality, safety, scheduling, metrics, and lean action plans to staff, (3) a chemotherapy safety log that tracks turnaround time and safe delivery of the pharmacy order, and (4) a just-in-time (JIT) inventory management system for intravenous supplies.
Results	Reduction of lab draw-time by 60 percent. Reduction of chemotherapy turnaround time from sixty-one minutes to forty-seven, a 23 percent improvement. The number of process steps decreased from thirty-two to sixteen, a 50 percent improvement. Increase of 30 percent in a number of chemotherapies. Face-to-face contact with patient increased by 50 percent.
Sector or Type of Organization	*Education:* Small private middle school in a suburban metropolitan area
WAs or URs Addressed	*Flow of work, quality of service:* Loss of instructional time that led to underachievement of curriculum goals.
Root Causes and Solutions	Process improvement tools were utilized to address issue of underachievement of curriculum goals. The main cause leading to the identified problem was frequent interruptions of instructional time.
Results	Recovery of 120 hours of instructional time per teacher over nine-month school year, higher levels of staff involvement in planning and scheduling of school and team level activities, and more-comprehensive exposure and learning at the student level.
Sector or Type of Organization	*Social services:* Small nonprofit organization in metropolitan area
WAs or URs Addressed	*Information deficits/organization of work environment/adequacy of resources:* Medicaid clients were not signing in/out as required for billing purposes.
Root Causes and Solutions	Process improvement tools were utilized to address location of sign-in sheet and poor floor plan layout, lack of needed pens and pencils and clock to record time, and errors in recording.
Results	Reduction of labor hours spent on acquiring the missing data resulted in annual savings of 1,734 hours and $54,000.

Source: The data in row 1 are from "Office of Child Support Enforcement Casebuilding Backlog," by Erie County Government, 2008, Buffalo, NY: Author. Adapted with permission. Retrieved April 11, 2010 from http://wwww.erie.gov/exec/public/pdf/Office%20of%20Child%20Support%20Enforcement%20Case%20Backlog.pdf

Data in row 2 are from "Doing More with Less: Lean Thinking and Patient Safety in Health Care," by R. A. Porche, Jr. (Ed.), 2006, Oak Brook, IL: Joint Commission Resources. Copyright 2006 by the Joint Commission Resources. Adapted with permission.

Data in row 3 are from "Proven Results Where Results are Most Needed: Case Study 1," by Lean Education Enterprise, Inc. Copyright 2010 by Lean Education Enterprise Inc. Adapted with permission. Retrieved on April 16, 2010 from http://www.leaneducation.com/case-studies.html#study1

Data in row 4 are from "Six Sigma Project: Improve Efficiency of Attendance Verification Pros Program," by Spectrum Human Services, 2009, Orchard Park, NY: Author. Copyright 2010 by Spectrum Human Services. Adapted with permission.

Form 7.2. Template for the A3 Report

Title: Name of the issue, problem, or theme of the effort	**Owner & Date:** Identifies the person, department or agency who owns the problem or issue and the date of the latest revision of the A3 form.
I. Background: Establishes the business context and importance of the issue. Answers the question, Why are you talking about the issue or problem?	**5. Proposed Countermeasures:** Diagram of the proposed new process (i.e., the future state map[s]) that will improve the current condition. Also may include information about how the recommended countermeasures will affect the root cause(s) of the problem.
2. Current Condition: Describes what is currently known about the problem or issue. It may include • process maps of current state, • identification of what is not ideal in the process or system, or • extent of the problem, e.g., current performance measures.	**6. Plan:** Describes an action plan of who will do what, how they will do it, and when will they do it.
3. Goals or Targets: Identifies the desired outcomes of the process improvement effort, e.g., the future state with performance measures.	(see Plan table below)
4. Cause Analysis: Analyzes the situation to determine the underlying root cause(s) that have created the gap between the current state and the desired outcomes (i.e., future state). May use • 5-Whys or • Fishbone Diagram.	**7. Follow-up:** Creates a follow-up review or learning process and anticipates remaining issues. It may include • how and when to check, and • status notes, with date of check, results compared to predicted, lessons learned, and next steps for ongoing continuous quality improvement.

WHAT Objectives	**HOW** Action Steps	**WHO** Lead & Others	**WHEN** Timeline

Source: From "Managing to Learn: Using the A3 Management Process to Solve Problems, Gain Agreement, Mentor, and Lead," by J. Shook, 2008, Cambridge, MA: The Lean Enterprise Institute, Inc. Copyright 2008 by The Lean Enterprise Institute, Inc. Adapted with permission.

6. **Plan**: Describes an action plan of who will do what, how they will do it, and when will they do it in order to reach the goal.

7. **Follow-up**: Creates a follow-up review or learning process and anticipates remaining issues.

While the specific format and content of an A3 report may vary from one situation to another, the critical feature associated with its use is that it leads the users to think systematically about the issues faced and changes needed to resolve them. Specifically, this tool can guide your process improvement team through the stages of

1. observing reality and gathering data to document the current condition;
2. presenting factual information in a succinct way;
3. proposing changes to correct the problem or issue to reach a stated goal;
4. gaining agreement among team members and management; and
5. following up with a process of checking results and making modifications, as required.

In essence, it is a "powerful tool for problem-solving, making improvements, and getting things done" (Shook, 2008, p. 10). Also, see Appendix 7A at the end of this chapter for an actual use of the A3 report at the University of Michigan Health System. In this particular case, lean tools were used to improve the discharge follow-up appointment process. This process in its current state was completed after the patient was discharged from the hospital. The hospital viewed follow-up appointments as a cornerstone for the continuity of care for discharged patients; continuity of care can reduce the number of readmissions or emergency room visits. Faced with a high rate of no shows and cancelled follow-up appointments, the hospital piloted a lean project with a goal of decreasing the number of emergency room visits or readmissions by previously discharged patients. Based on this presenting problem, the improvement team devised an alternative process for scheduling follow-up appointments that was standardized and completed prior to patient discharge.

An important piece of information in this A3 report is with respect to the number of changes made to the process for scheduling follow-up appointments. The first attempt (i.e., "intervention," in their terminology) to modify this process was not successful, based on the data for no-shows and cancellations. However, another pilot in the emergency room discharge process was implemented that included the patient in scheduling discharge appointments and the development of an online appointment request tool. Results from this second intervention showed it to be effective in reducing the number of cancelled appointments and no-shows.

Therefore, not only did this improvement team have multiple attempts at a process redesign, but also they were diligent in collecting the outcome or results data, which was used to determine the effectiveness of the process change. Based on the examination of outcome data from the two pilots, the second process was adopted and the hospital is now determining where this improved process can be implemented in other areas of the hospital. This example of an A3 report also provides key information about lessons learned that will be instrumental as improvements in the discharge process and scheduling of follow-up visits are made elsewhere.

> *You don't have to see the whole staircase to take the first step.*
>
> — Martin Luther King, Jr.

Appendix 7A

FIGURE 7.3. University of Michigan Health System A3 Report[1]

University of Michigan Health System

Improving Patient Flow by Reducing

Name of Project or Theme: LEAN Discharge Follow up Appointment Process

Brief History

Each year, the University of Michigan Health System (UMHS) treats more than one million outpatients, provides at least 36,000 hospital visits, conducts hundreds of scientific research projects and educates the next generation of medical professionals. UMHS has experienced high occupancy for the past 12 months with an average occupancy of 93%. In an effort to improve patient flow and increase capacity a lean project was commissioned to study the inpatient discharge process on a pilot unit. The time after discharge is considered high risk for patient care, often marked in our patients by re-admissions and/or repeat ER visits. The follow-up appointment from discharge is hypothesized to be the cornerstone of continuity of care to prevent re-admits/ER visits.

In 2006, prior to an intervention, 48,954 discharge follow-up appointments were scheduled primarily after the patient was discharged. Approximately 60% of patients arrived to their appointment, 15% were no-shows and 25% cancelled. A 1st intervention on the Medical Faculty Hospitalist Service (MFH) was piloted, scheduling appointments prior to discharge. Due to process issues, this change did not significantly affect the rate of no-shows and cancellations. A separate ER pilot had been implemented to improve follow-up appointments from ER discharges. For the 2nd intervention, this process was adapted to include the patient in scheduling discharge appointments and online appointment request tool was developed and piloted.

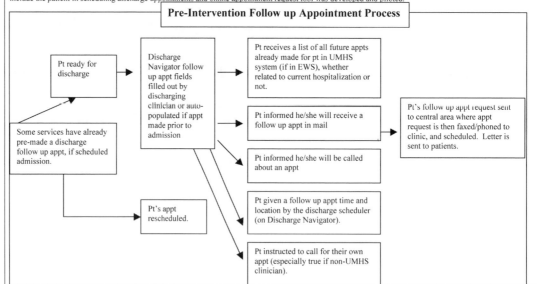

Pre-Intervention Follow up Appointment Process

Future State Goals: Ensure that patients have a communicated follow up appointment(s) at the time of the discharge from the hospital, to promote a smooth transition of care to the outpatient setting. A specific goal for the discharging physician is to focus on determining which follow up appointments are important as related to this hospitalization. An anticipated outcome was that emergency room visits and/or readmissions would decrease.

- Appointment made prior to patient leaving, at least 24 hrs prior to discharge

- Appointment made with patient/family involvement

- U of M Attending Physician is notified that appointment is made

- Nursing includes follow up appointment information during discharge instructions

Who Is Involved? (Major stakeholders and players)
Lean Process Owners: Robert Chang, MD and Donna McClish, RN **Lean Coaches**: Kate Bombach and Christopher Kim, MD, **Public Relations and Marketing Communications:** Josie Aguirre and Laura Rowland **Outpatient clinics staff, Patients and Families**

Countermeasure Implemented to Date:

- A web site was developed to allow physicians to place requests for discharge follow-up appointments for patients on the 5B inpatient unit. (See attachment 1)

- The workflow process for discharge follow-up appointments was standardized. (See attachment 2)

  o   U-M physician completes Web form to request a discharge follow-up appointment with either the patient's PCP or a specialist.

  o   A discharge appointment coordinator receives the request in a work queue. The coordinator contacts the patient directly, while still in the hospital, and reviews the discharge instructions for a recommended follow-up appointment. (See attachment 3)

  o   The discharge appointment coordinator connects the patient to either the PCP's office or specialty clinic by telephone to schedule a follow-up appointment within the suggested time frame.

- The process allowed the nurse to review the online discharge appointments in Discharge Navigator.

  o   As part of the discharge process, the unit nurse reviews any necessary patient education information and future scheduled appointments.

- An easy online process for tracking the appointment outcome was developed for the U-M physician

  o   The discharge appointment coordinator documents the appointment outcome in the Discharge Follow-Up Appointment Web application, for the U-M physician to track.

Hospital Readmissions through Patient Involvement

Where Are We in the Project Cycle? Full rollout to MFH Service - October 2007

Discharge Follow Up Appointments Workflow (June 2007)

INTERVENTIONS

- Inpatient Physician completes web form and electronically submits to Discharge Appointment Line
- Patient information is reviewed in Careweb/Mainframe: insurance, PCP, inpatient phone number
- Contact patient and explain reason for call
- Is patient available to schedule appointment? NO — Identify best time to call patient again. YES — Explain scheduling process for the type of appointment(s) needed.
- Connect to clinic. Before connecting patient to scheduler provide: Patient Name / CPI / Diagnosis / Ordering Physician / Time frame
- Image* radiology requisition into Careweb under "Imaged Docs" tab
- Update radiology requisition with appointment date, time and location
- Update Discharge Appointment Application with appointment information
- Update Discharge Navigator with appointment information
- File Referral

Discharge Follow Up Appointment Request (University of Michigan Health System web form with fields: Submitter's name, Pager number, UMHS Registration Number, Patient Information, Referral to PCP, Specialty Clinic Referral #1 and #2, Diagnosis & Reason for Consult, Referral to Diagnostic Services, Special Needs, Submit Your Request)

Appointment Outcome Data

Appointment Outcome

Pre-intervention
(12/1/2004-11/30/2005)
- Scheduled Appts. 398
- Arrived 237 (60%)
- No Show 61 (15%)
- Cancelled 100 (25%)

MFH Full Rollout
(9/6/2007 – 11/9/2007)
- Scheduled Appts. 249
- Arrived 178 (72%)
- No Show 20 (8%)
- Cancelled 51 (20%)

Readmission Rate < 14 days

Pre-intervention 11.81%

MFH Full Rollout 7.90%

Return to ED < 3 days

Pre-intervention 3.96%

MFH Full Rollout 0.75%

Lessons Learned:
- Multidisciplinary team consisting of physicians, nurses, discharge planning, information technology, public relations designed an application that was very user friendly.
- The Michigan Quality System and the lean Healthcare Method was an effective methodology. Small scale testing led to rapid improvements in the Web forms.
- Understanding physician workflow and tools can improve work flow.
- Take advantage of pre-existing technology and expand on it. The inclusion of individuals from Information Technology is essential.
- Frequent, focused and succinct communication was key. This included soliciting feedback from the faculty and staff.
- Involvement of patients and families was essential.
- Standardized training for the physicians is essential. In order to be effective, the training needs to be focused and hands on.
- Multiple rounds of testing helped to flush out problems and user interface design.

Next Steps:
- Patient satisfaction data collection
- Firming up takt time: number of calls the scheduler can handle per day and how long each call takes
- Discussion of expansion after stress test is over and above steps completed
- Hospital-wide – will look at other services (i.e. surgery often schedules their f/u appointment in clinic prior to admission)

Appendix 7B

FIGURE 7.4. Action Plans Subsidy Program

Name of program/service: Subsidy Program	**Date:** January 25, 2007

Process mapped: Completion and return of Childcare Arrangement form (DCC-552).

Goal of process improvement: To ensure the accurate completion and proper return of the Childcare Arrangement form (DCC-522) to the designated case worker

UR addressed: Rework: inaccurate forms, incorrect authorization worksheets

WHAT: Objective*	**HOW:** Action Steps*	**WHO:**		**WHEN:** Timeline
		Lead	**Other Stakeholders or Partners**	
1. To ensure the Child-care Arrangement form is completed accurately by provider and given to parent to submit to assigned case worker for accurate authorization assignment.	1-a. Develop a checklist and update the Childcare Arrangement form to make it easier to complete accurately.	Tonji		July 1, 2007
	1-b. Inform providers and parents of the new process for completing the Childcare Authorization form.	Margaret		July 1, 2007
	1-c. Distribute forms to providers.		Doris (print request) Margaret	July 1, 2007
	1-d. Update information and checklist on the Child-care Arrangement form.	Nancy		February 2, 2007

STATUS NOTES: 1. To ensure the Childcare Arrangement form is completed accurately by provider, given to parent to submit to assigned case worker for accurate authorization assignment.

Date	**Comments**
April 3, 2007	Ivory made some changes on the Childcare Arrangement form that will be sent to the provider who will fill it out.
	Question about whether the form should still be part of the parent packet or just distributed to providers who fill it out and give to the parent to return with their completed application.
	Also, suggestion was to have a checklist, which tells the applicant what has to be turned in with the application.
	Kristy raised question about not having the ABC option for childcare needs. That option currently is not noted on the form.
	Suggestion made by Margaret to duplex first two pages.
	Still need to decide whether or not to include the CCA form in the parent packet.

STATUS NOTES: 1. To ensure that the Childcare Arrangement form is completed accurately by provider and given to parent to submit to assigned case worker for accurate authorization assignment.

Date	**Comments**
May 24, 2007	This is a section of the DCC-522 and it has to be promulgated before it can go through piloting. First thing is public hearing, scheduled for June 29 (new application is part of new policy): Notice goes to public in public paper (three days prior to hearing); plan to have the notice in newspaper along with DHHS website and newsletter. Input from hearing may result in changes (not likely). You can set start date whenever Ivory decides. Right now proposed start-up date is August 1 or September 1. Subsequent to this, it goes through the following process: (1) Draft copy goes to Division to determine if it is legally corrected (generally takes one to two days). (2) Have paperwork filled out and take it with paperwork to Bureau of Legislative Research (wait three days). (3) Pick up approved copy that goes to AR State Library and secretary of state for their records. (If legislative body is in, it could take longer since Bureau of Legislative Research will have other work to do; therefore, it could take up to six weeks.)
	Question was raised about whether the form is ready to go through the promulgation process—there was some confusion on the terminology of "casehead." Suggestion is to add in parentheses "Parent/Guardian"
	Next steps to pretest with small group before taking to public hearing. But because of this process, there will not be a pilot as originally designed. Therefore Ivory will do smaller pilots prior to the public hearing (have some parents fill it out to ensure they understand it and then have it go through the internal process to ensure it will work).

From Knowledge to Practice

Part III presents a number of lean tools and methods you can use to implement a lean transformation within your organization. The following exercises provide you with the opportunity to take the knowledge gained in this section and put it into practice.

Exercise III.1. Mapping a Value Stream
Exercise III.2. Creating a Process Flow Map for a Macro Step
Exercise III.3. Completing a Root Cause Analysis
Exercise III.4. Preparing an Action Plan to Improve a Process
Exercise III.5. Tracking Performance Measures Over Time

In light of our emphasis on learning by doing, the final activity in Part III asks you to reflect on the exercises completed and discuss the challenges associated with each of the exercises and any lessons learned as a result of your efforts that can help you in further attempts to apply this knowledge.

Exercise III.1. Mapping a Value Stream

Based on the exercises you completed in Part II, take a value stream that has been identified as one filled with WAs and URs. For this value stream, complete the following:

1. Name of program or service:

_____

2. Value stream or process to be mapped:

_____

3. Responses to these four preparatory questions:
   • What is the purpose of the process? _____

   • What are the indicators of success? _____

   • What are the specific performance measures? _____

   • Who is responsible for the process? _____

4. Identify your core team of individuals to complete the mapping and analysis activities (for this exercise, identify at a minimum three individuals, including yourself):

_____

_____

5. With your core team and use of the appropriate data gathering methods identified in Chapter 5, create the VSM, including the following components:
   • macro steps,
   • information flow, and
   • key performance data describing the process's current state operation: total cycle time, VA time, number of individuals, waiting time, elapsed time, and dollar value of supplies or materials (if applicable).

Calculate percentage of total elapsed time that is VA: _____

Exercise III.2. Creating a Process Flow Map for a Macro Step

Select one of the macro steps of the VSM created in Exercise III.1 and create a process flow map, using the icons identified in Table 5.3. Once the process is mapped, with your core team, complete the following:

1. Determine for each task or step in the process flow map the amount of time it takes to complete the task. Also, identify the total elapsed time from when the process step begins to when it ends.

2. Assess each task or step and label it as either (a) value-added (VA), (b) non-value-added (NVA), or (c) required non-value-added (RNVA).

3. Based on your determination as to whether a task or step is VA, NVA, or RNVA, complete a Process Flow Map data collection form, as found in Appendix 5B.

4. What percentage of time in this process is NVA? _____

5. Discuss what this tells you with respect to the opportunities to improve the operational efficiency or effectiveness of the process.

Exercise III.3. Completing a Root Cause Analysis

1. Based on the mapping and data gathered regarding areas of WAs and URs, select one problem area and complete a root cause analysis using either the 5-Whys or the Fishbone Diagram tool.

2. With your core team, take your root cause analysis and brainstorm possible solutions to the problem that will eliminate areas of waste and URs associated with the current state process. Prepare a Benefit vs. Cost/Time Matrix for these solutions. (See Figure 6.4.)

3. Take at least one of your solutions that uses any of the lean tools discussed in Chapter 6 and explain how it applies to 5S, workload balancing, or visual controls and management.

Exercise III.4. Preparing an Action Plan to Improve a Process

1. Using the Action Planning Tool discussed in Chapter 7 (see Form 7.1), develop an action plan to implement at least one of the proposed solutions identified in Exercise III.3.

Exercise III.5. Tracking Performance Measures Over Time

1. Using the performance measures that you identified in Exercise III.1, and specifying hypothetical data for the after measure, prepare a chart or graph that shows the change in the performance measure from before to after your improvement was implemented.

2. Provide a discussion as to the implications of this data analysis with respect to how it would impact organizational learning and the next process improvement steps to take.

Reflections and Lessons Learned

1. Based on your completion of the exercises in Part III, reflect on both the challenges and lessons learned. Provide a discussion of these challenges and lessons learned. Also, identify specific ways that you can make improvements the next time you use these lean tools to complete mapping, analysis, action planning, and tracking of results.

Sustaining Improvements Over Time

The previous parts of this book laid the foundation for improving performance in your service organizations based on the application of lean philosophy. The concepts and methodological tools that have been introduced offer an approach to improving your performance through the analysis of processes to identify areas of waste and the unacceptable results that are produced as a function of the way processes are designed and implemented. Hence, the design of processes and how they operate greatly impact your organization's optimal efficiency and effectiveness.

Part IV of this book turns to a discussion of the factors that can either enhance or inhibit your organization's ability to sustain process improvements efforts over time and offers an in-depth view of some lean transformations in service-based organizations. Chapter 8 begins this discussion by providing an overview of a number of challenges and pitfalls associated with continuous quality improvement efforts, particularly those associated with lean transformations. In keeping with our position that challenges represent opportunities for improvements, this chapter identifies the opportunities to redesign processes to produce significant improvements in your organization's performance.

The second part of Chapter 8 goes a step farther and describes a number of strategies that you can use to create a culture and climate within your organization where the notions of continuous quality improvement and organizational learning are embedded within your norms, values, and expected ways of behaving. After examining the culture, attitude, and behavior relationship, this chapter identifies the critical factors within an organization that can predict successful lean transformations. Furthermore, it provides an overview of the type of lean management system that is instrumental in sustaining improvements and ensuring that workers are continuously looking for ways to eliminate waste, problem solve, and design processes that are more efficient and effective.

The final chapter in this book, Chapter 9, offers three different case studies of lean transformations in service organizations. The cases provide greater insight into how lean concepts and methods can be applied in organizations that provide services and other material deliverables to clients, as opposed to organizations that produce a product in a manufacturing work environment. Cases are presented for education, government, and social service work environments.

CHAPTER 8

Creating a Culture of Organizational Learning and Your Lean Transformation

We begin this chapter by taking a look at the relationship between culture, attitude, and behavior, and what that relationship means with respect to creating an organizational culture and climate within your organization that supports ongoing organizational learning. Next, we identify challenges your organization may face in its lean transformation; we will further examine a number of these challenges in our discussion of the critical factors in creating a lean organization. We conclude this chapter with a fuller discussion of the role of leadership, the role of employees, and your management system, all of which are critical factors in determining the success of your lean transformation. There is some overlap in our discussion throughout this chapter, which is unavoidable. When you take steps to create a management system to support your transformation efforts, you are addressing the interrelationship between your organization's leaders and employees, and dealing with the challenges of establishing the right conditions, organizational structures, and normative system (i.e., culture and climate) that will support your lean transformation efforts.

ORGANIZATIONAL CHANGE

Undoubtedly, organizational change can be challenging regardless of the philosophy underlying it. There are a number of significant challenges you may face in your effort to implement improvements (particularly lean transformations), sustain them over time, and engage in a cycle of continuous change and learning. The research and commentary on organizational change and the factors that facilitate it seem to be unlimited; one could spend countless hours synthesizing them. We begin with an examination of the relationship between culture, attitudes, and behavior; this lays the foundation for understanding organizational change, determining the best strategies for establishing a culture of organizational learning, and addressing the challenges organizations face when undergoing a lean transformation.

Culture–Attitude–Behavior Relationship

How can you create an organizational culture where the notions of change and improvement are embedded within your shared organizational norms, values, and expected ways of behaving (culture), and where those same notions are reflected in the beliefs, perceptions, and behavior of individual organizational members on a daily basis (climate)? Without a doubt, your organization must establish a culture and climate that supports continuous quality improvement and organizational learning, because, without it, your use of lean tools to improve processes will not be sustained over time. Lean thinking as a way of doing work must become a habit, without which process improvements are likely to be short-lived, with behaviors reverting to the old way of doing things.

To understand how you can create a learning culture in your organization, we need to examine the relationship between organizational culture and the attitudes and behaviors of individuals within your organization. We are all familiar with the classic question about which came first, the chicken or the egg. In any environment, whether the level of organization is at the micro level (such as a family), or at the macro level (e.g., small organizations all the way through whole societies), there is a complex interrelationship among the behavior of individuals, the attitudes those individuals hold, and the culture within which they exist.

A typical way of thinking about this interrelationship is that culture sets the standards or rules that govern behavior—e.g., the *do's* and *don'ts* of how to behave (norms) and the values associated with these norms: individuals will behave according to established cultural norms and values. If only it were that simple. While we know it is not that simple, there is still a reliance on first establishing a new culture by creating a new set of rules and educating people about the norms and values that are associated with this new culture. With this new culture in place, it is expected that people will change their attitudes and beliefs about what is important and subsequently change their behavior so that it is congruent with the established culture.

However, as has been evidenced in practice, it is essential (1) to create an environmental structure first that will drive behavior in a particular way, and (2) to have mechanisms in place to reinforce behavior over time. Taking these two steps eventually will lead to a new culture that has established a new normative system and the values that go along with it. Therefore, when beginning a lean transformation within your organization, you need to introduce these tools and implement a system of reinforcement for

> *We are what we repeatedly do. Excellence, then, is not an act, but a habit.*
>
> — Aristotle

their repeated use. These elements (introduction of tools, use of tools, and reinforcement) will eventually result in a cultural change where your entire organization "begins to change radically from *catching mistakes* to *preventing mistakes*" (Tapping & Dunn, 2006, p. 13; emphasis in original).

CHALLENGES OF A LEAN TRANSFORMATION

Having this theoretical understanding of the relationship among culture, attitudes, and behavior does not make the cultural change process easy. There is no magic bullet that can transform your organization's culture and climate overnight. It takes considerable thought and discipline to create new habits, as we are all well aware: for instance, think about how much thought and discipline it takes to extinguish bad habits (e.g., smoking, unhealthy eating, not exercising regularly, etc.). It takes dedication, diligence, and hard work to make a transformation a reality over time. In an ideal world, our attitudes, behavior, cultural norms, and values are all aligned. We recognize the importance of striving for excellence and our efforts to improve the way we do our work are wholeheartedly embraced. Unfortunately, we do not live in an ideal world, and while change is the one thing in life we can always count on, it does not make it any easier to handle, even if it is focused on efforts that will make our lives better. Fear of the unknown and resistance to change are powerful emotions that are difficult to counteract, regardless of the model of continuous improvement used. Given this, it is important to have a clear understanding of the challenges and pitfalls to avoid when implementing improvement efforts, particularly lean transformations within service organizations.

> *Bad habits are like a comfortable bed: easy to get into, but hard to get out of.*
>
> — Anonymous

We have identified the top ten challenges and pitfalls that we categorize as (1) attitude and commitment challenges, (2) challenges with unrealistic assumptions and goals, or (3) execution challenges. Some of the solutions are similar across one or more of the challenges, hence the challenges are interrelated and not mutually exclusive of one another. This discussion provides you with a knowledge base and understanding of the challenges that are instrumental for identifying the opportunities to design a system of organizational learning based on lean philosophy, a system that will produce significant results with respect to an organization's performance. A summary of these challenges is provided in Figure 8.1.

FIGURE 8.1. Ten Challenges to a Lean Transformation

Challenge	Solutions to Address Challenge
Attitude and Commitment Challenges	
Challenge I: *Lack of executive sponsorship and commitment*	• Include top-level management in an integral way in the process improvement efforts.
	• Clearly communicate the potential value of improvement efforts in the language of senior management, e.g., bottom line, survival, or mandates for accountability.
	• Remain diligent in your implementation process and continuously engage in the Plan Do Check Act (PDCA) cycle until you reach your goals.
	• Maintain sufficient controls to sustain the effort over time, preventing behavior from reverting to previous states.
	• Document successes of improvement efforts and communicate results throughout your organization.

FIGURE 8.1. Ten Challenges to a Lean Transformation—(*Continued*)

Challenge	Solutions to Address Challenge
Challenge II: Difficulties in engaging employee participation	• Take very clear steps from the beginning to engage employees in identifying problems or issues with the way work is done. • Involve employees in problem-solving activities and identifying solutions to process issues.
Unrealistic Assumptions Challenges	
Challenge III: The "quick fix" lean system	• Establish an organizational culture and climate where leaders and employees embrace ongoing organizational learning as the norm. • Accept and encourage routine creativity and problem-solving at *all levels* of the organization.
Challenge IV: Inadequate level of lean understanding	• Send a clear message that you learn by doing, that the process of learning is ongoing, and that lean transformations are hard work. • Do not include completion dates on process improvement efforts to support the notion that improvement is ongoing. • Have ongoing recognition and celebration of successes. • Always include next steps when reporting improvement efforts and results, even if it means taking the learning and applying it to another work process.
Challenge V: Taking on too many initiatives at once	• Make sure the team leader has the experience, knowledge, and skill set to facilitate group processes, motivate team members, trouble-shoot when necessary, and keep everyone on track with action plan implementation.
Challenge VI: Being clear on process and outcomes of service organizations	• Recognize and accept that value streams and performance measures will not be perfect and may be established through observation or educated estimations provided by those who implement a process. • Stabilize a process before process improvement efforts are implemented or as part of the improvement process. • Start improvement efforts with the low-hanging fruit as soon as possible to reinforce the positive results that can happen when lean solutions are applied.
Challenge VII: Selecting improvement projects not aligned with organizational goals	• Clearly specify organizational goals using a logic model or strategic planning process before undertaking process improvement efforts. • Identify and rank order projects as to their relevance and impact on a particular work area, and as to their relevance and impact for accomplishing organizational goals.
Execution Challenges	
Challenge VIII: Inappropriate improvement team membership	• Determine the groups that should be represented on an improvement team. • Identify full-time team members (no more than six to ten) to represent key players and add supporting members as necessary. • Clearly document team efforts and outcomes and distribute information or brief everyone about the improvement efforts.
Challenge IX: Ineffective control plans	• Do not make work more difficult and cumbersome by establishing more instructions, signoffs, and auditing procedures to control a new process. • Carefully consider and use ingenuity of process improvement team and frontline workers to ensure that new processes (1) are easier and more enjoyable to run than the old, (2) do not have unacceptable results that would make work frustrating and undesirable, and (3) make it difficult to go back to the old way of doing work.
Challenge X: Inappropriate management of process improvement efforts	• Do not change the process improvement effort by jumping to solutions too quickly or falling short of completing the full cycle of planning, implementing, evaluating, and revising. • Select the right person to be the team leader and the right persons to be coordinators at the work areas; they will have (1) experience in the service area, (2) knowledge of tools and techniques of lean thinking, and (3) soft skills to put the experience and knowledge to work.

The greatest discovery of my generation is that a human being can alter his life by altering his attitudes of mind.

— William James

Attitude and Commitment Challenges

Without a doubt, the top two challenges in a lean transformation are attitude and commitment challenges of your organization's leaders and employees, as discussed below (challenges I and II).

Challenge I: Lack of Executive Sponsorship and Commitment. Successful lean transformations require support of your upper-level management for process improvement efforts, a level of support that translates into a passionate commitment to excellence and leadership that is effective in communicating your message throughout an organization. Without this commitment and support, even the most diligent improvement efforts from your middle management or the frontline workers will lose steam, particularly if the improvement team must continuously beg for needed resources or validate what they are trying to do. If this type of climate exists within an organization, it will not take long for employees to become frustrated, cynical, or—worse—apathetic about their work.

There may be many reasons for what appears to be a lack of sponsorship and commitment by your upper-level management (Riley, 2008): (1) a misunderstanding of the potential value of improvement efforts, (2) a poor implementation process that results in missteps and endeavors that do not produce intended results, (3) insufficient controls to sustain the effort over time resulting in behavior that reverts to previous states, (4) inadequate validation efforts to document successes, (5) loss of focus on the bottom line, since financial security is a major concern of management. As Mann (2009, p. 15) has reinforced in his discussion of lean leadership, "implementing [lean] tools represents at most 20 percent of the effort in lean transformations. The other 80 percent of the effort is expended on changing leaders' practices and behaviors, and ultimately their mindset."

To ensure that top-level management maintains a commitment to your lean transformation, if the transformation is driven by middle management it is essential that when implementing your improvement efforts you

1. include top-level management in an integral way in the process improvement efforts;
2. clearly communicate the potential value of improvement efforts, particularly as they relate to results associated with your organization's fiscal goals, survival, or its mandates for accountability that make up the language of senior management;
3. remain diligent in your implementation process so if your endeavors do not produce intended results, then you continuously engage in the **P**lan, **D**o, **C**heck (or Study), and **A**ct (PDCA) cycle until you reach your goals;
4. maintain sufficient controls to sustain the effort over time, preventing behavior from reverting to previous states; and
5. document successes of your improvement efforts and communicate these results throughout your organization.

Challenge II: Difficulties in Engaging Employee Participation. The flip side of executive support and commitment is getting the buy-in of your frontline workers in the design and implementation of your lean transformation. Getting that buy-in probably is one of the biggest challenges. There are examples of lean transformations that moved forward ruthlessly, with massive reengineering of companies resulting in significant job loss that relied on the command-and-control approach to manage the transformation (Mitchell, 2000). Getting buy-in is a challenge because there have been examples of ruthless transformations with job loss. This leads to distrust among front-line workers. In addition, there may be distrust between management and line-staff if your organization has a culture of "killing the messenger" when problems are raised about how work is done. Finally, your employees may be set in their traditions and be resistant to any change that disrupts their routines, or they may think, "How could I, a talented employee, create something bad—i.e., wasteful?"

Lean transformations often fail because of an inability to manage the people component. If your lean initiatives are driven by management groups who decide the efforts, pick the people to put on teams, and host the events, then these improvements are management driven, not employee driven. The Achilles heel of kaizen efforts "is that the process is done around people, not with people" (Butler & Snyder, 2001, p. 6). This approach discourages independence and therefore forfeits the employee ownership that is necessary for sustainment. Your organization's lean program will be seen as another "flavor of the month" to be tolerated, since it will go away eventually.

It is essential to garner workers' cooperation, as well as their involvement in the change process. If that does not happen, efforts to change behavior associated with the operation of new processes will likely fail in the long run or there will be a great deal of resistance to change efforts. Unfortunately, "going lean" often has been translated as "lean and mean" or "fewer employees will be needed." It is natural for individuals to think about "what's in it for me" with respect to efforts that propose to improve operational efficiency. Given this, your employees will be attuned to messages (whether intended or not) that have a hint of labor reductions. Moreover, there is a human tendency to reject any notions that one's work is full of waste, particularly in situations where employees have greater latitude in developing their work habits and processes; this is more often the case in service organizations or in processes that provide a service rather than it is on an assembly line.

As a way to categorize the levels of support for change efforts, Tapping and Dunn (2006) describe these types of individuals and their distribution within an organization:

1. **The Doers**: These are individuals who are instrumental in making change happen within an organization.
2. **The Helpers**: These are individuals who will help and provide assistance in the change effort.
3. **The Followers**: These are individuals who will let the change happen and not resist it.
4. **The Resisters**: These are individuals who are mildly against the proposed change.
5. **The Saboteurs**: These are individuals who will actively oppose change efforts and engage in activities to sabotage it.

According to Tapping and Dunn, an "80 or 20" rule can be applied here: no more than 20 percent of an organization's employees will fall into the last two categories of resisters or saboteurs. The other 80 percent are supporters to some degree or another. In addition, approximately 80 percent of the resisters and saboteurs can be converted into supporters. Given this rule, it is important for you to find ways to facilitate this conversion and to focus efforts on those who are supportive of the improvement efforts.

To ensure worker buy-in, you must take very clear steps from the beginning to get workers engaged in identifying what is important to them, especially with regard to problems or issues they have with the way work is done; you must then involve them in the problem-solving activities. If your employees are convinced that a lean transformation

> *No pessimist ever discovered the secrets of the stars, or sailed to an uncharted land, or opened a new heaven to the human spirit.*
>
> — Helen Keller

will eliminate many of the hassles they currently contend with in their work, they are more likely to have the motivation to make the necessary changes to achieve that result. Additional guidance on how to engage employees will be presented later in this chapter under our discussion of the role of employees in your organization's lean transformation. As an example of how one service organization was able to secure buy-in from employees and successfully implement process improvements, see Example 8.1.

EXAMPLE 8.1. Eflexgroup.com's Road to Achieving Full Organizational Buy-In for Lean Six Sigma

If ever a company needed the tools in the Lean Six Sigma toolkit, it was eflexgroup.com (eflex). Based in Madison, Wisconsin, we are a third-party administrator of health flexible spending accounts, dependent care accounts, HSAs, etc., for employers across the country. Our advertising includes three phrases: Fast Claims, Fast Answers, and Web Self-Service. Therein lay our problem: Our claim processing was not meeting the 24-hour turnaround we promised. In fact, they were averaging from seven to 10 days in early 2007. That problem would have been enough, but the long processing times were causing up to 2,000 phone calls a day. Customers were leaving us, and we needed to do something now.

Here's a Book on Lean Six Sigma, Now Let's Stop the Bleeding
It started with an invitation from our CEO, Ric Joyner, to a meeting in our conference room. Four of us from all different parts of the organization, customer-facing and not, were invited. Ric came in and asked us to take on the duty of being the Quality Team for eflex. We all agreed to be part of the team, as we were all passionate about our company and about excellence.

The next thing out of the CEO's mouth was that we were going to stop selling actively. We needed to stop the bleeding in claims processing before we sold any more because we were not delivering our value proposition—paying claims quickly. The tool we would use to solve the problem and stop the bleeding, Ric said, was Lean Six Sigma. Ric distributed the book *Lean Six Sigma for Service* by Michael George to each and every one of us.

No Flavor of the Month Here: How We Achieved Lean Six Sigma Buy-In
The moment he left the room people started commenting, saying, "He's taking another class again, isn't he?" and "Great, another addition to the managers' book of the month club." It was not met with excitement. But, confession is good for the soul: Secretly I loved this stuff!

From the start we faced challenges. There was not a single trained Lean Six Sigma Belt of any level. Ric only knew of Lean Six Sigma from a chapter in an operations textbook from his MBA studies. Consultants were not an option financially, and there was no outcry of enthusiasm for the decision. But we did have something critical: We had C-suite buy-in. Much has been written about how critical this element is to the success of a program.

Ric put his money where his mouth was and let everyone in the company know that this was how we would do business from that point on. To that end, as we worked on the project, he searched out online training for our company and paid for it for anyone interested in putting in the effort. He did not allow Lean Six Sigma to become a flavor of the month.

Our first conversations with the processors in claims, our frontline workers, were not breakthrough moments. The original responses were similar to those heard in businesses around the world: "This is the way we have always done it," and "The process is working just fine." Our processors were afraid for their jobs. Some had actually heard of Lean Six Sigma and had the idea that this was about getting rid of bodies. These are not easy to overcome; just read one of the many online discussions of them!

To start with, both our CEO and our president told not only the claims department, but the entire company, that our adoption of Lean Six Sigma was not about reducing headcount. Normally this reassurance might not be enough, but our company was smaller and there were direct connections between the CEO, president and frontline workers. Our workers trust our leadership, so they believed it when they were told that it was not about getting rid of bodies. By the same token, Ric was blunt in stating that there was a problem. He put people on notice that they could be part of the solution or be, at best, re-assigned.

At the same time, he and the Quality Team listened to those claims processors. Our workforce is not just here in Madison, but throughout the country, so the listening sessions were scheduled to fit the time zone differences and the schedules of those who work at night. They told us what was *really* happening within the process and had ideas about how it could be better. And they lived the company value of taking care of our customers.

Better than just listening to the workers who do the work, we took their ideas and used them. We redesigned the process and then let them pick it apart. When they told us that we would need to retrain our workers, we did. When they told us "it would take less time if . . . ," we changed it. They ended up helping us to rewrite the job descriptions of these workers.

Assessing How We Got It Right

So here is a critical idea—we did not start with buy-in for Lean Six Sigma; we had to earn buy-in. How? Management commitment to not reducing headcount; calling on the trust built over time between management and the frontline workers; not just having listening sessions with frontline workers, but changing things they said needed changing, thus building more trust, all of these were critical to our success. Top-down deployment only went as far as saying that we were going to use the Lean Six Sigma methods and tools, it did not prescribe the way that any business process would change.

Our staff, and yours too I suspect, *do* talk with each other. Leveraging that first success, reducing claim cycle time to under a day, was not only about using the monetary savings or the cycle time reduction to gain Lean Six Sigma buy-in throughout the company. Imagine what kind of momentum and acceptance we would have had if we had reduced headcount, if we had listened to suggestions but not implemented any of them. Word would have gone out immediately that this was just management telling us that they know better and ideas from others are not needed. And our next Lean Six Sigma project would have died before it was chartered. How is your company, your deployment doing along these lines?

eflex Continues to Sustain the Momentum for Lean Six Sigma

We did get mileage from the results of that Lean Six Sigma project. To that fuel we added continued executive engagement— our president comes to the kickoff meeting of every project. He has read the pre-Define charter and knows what is happening. When we get to pitching our pilots for improvement, he is there, and so far has allowed at least initial exploration of every Lean Six Sigma project presented to him. He does not want anything to dampen the creative impulses of a project team.

More successful Lean Six Sigma projects combined with the consistency of message from our executive leadership continues to propel us forward. We communicate the results of our projects across the company. As a result I now have people from all departments coming to me and talking about opportunities in their departments. In fact, our sales department is now engaged in a large Lean Six Sigma project right now. They have a Green Belt in-training who will deliver a project on another aspect of sales later this spring. There are even people out there who *want* to take Lean Six Sigma training because of their experience in a project.

The Voice of the Customer drives us, but that voice includes our clients, the voices of members of our Lean Six Sigma project teams, department heads and all the rest of our employees. That is what is bringing us success.

Source: From "Case Study: eflexgroup.com's Road to Achieving Full Organizational Buy-In for Lean Six Sigma," by J. Cox, 2010. Copyright 2010 International Quality and Productivity Center. Reprinted with Permission. Retrieved April 16, 2010 from http://www.sixsigmaiq.com/sponsor_article .cfm?externalID=1982

Unrealistic Assumptions Challenges

There are five unrealistic assumptions and goals that can derail a lean transformation effort: challenges III to VII:

Challenge III: The "Quick Fix" Lean System. If your organization is faced with operational challenges, particularly those that may jeopardize its survival, you may be driven to find immediate solutions. The downside of immediate solutions are half-hearted attempts to implement a quick fix lean system that is management controlled, whose driving focuses are reducing costs and increasing revenue. Too often, these attempts have a shortsighted view of lean, with only sporadic lean projects (i.e., kaizen events) implemented, devoid of any long-term lean strategy (Butcher, 2007). As mentioned earlier, without long-term strategic thinking about process improvement efforts and how to sustain them, your sporadic efforts become just another flavor-of-the-month that workers see come and go. In this case, it is not unusual for your workers to make half-hearted attempts to implement improvements and give lip service to them, all the while saying that if they hold out long enough, they can revert to their old way of doing their work.

To avoid the "quick fix" syndrome, it is vitally important for your organization to establish a habit where your leaders and employees embrace organizational learning over the long term and that you have a climate that accepts and encourages routine creativity and problem-solving at all levels of your organization. This was also true of Challenge I and Challenge II.

Challenge IV: Inadequate Level of Lean Understanding. Undoubtedly, both leadership commitment and employee buy-in are critical factors for successful lean transformations; however, the essential ingredient for long-term success has to do with the frame of mind of both workers and management. This means that one should (1) always assume there is more to learn, (2) recognize that there is no end to improvement, (3) reject the notion that complete understanding or excellence has been achieved, and (4) embrace a way of life that accepts challenges or problems as opportunities for learning and improving, rather than as failures or barriers to success.

If your organization sets a level of lean education, saying that the number of hours of training, certifications, or quantity of readings completed are enough, then you are sending the wrong message to your employees. This message establishes a line between learning and doing, and ignores the idea that one learns by doing and that the learning process is ongoing.

Another aspect of improvement planning and lean transformations that can send the wrong message is if you establish a date when a lean effort will be completed. Hagood stated (as quoted in Butcher, 2007, p. 3):

> A lean transformation has no end date. The process is ongoing and is never a closed-out action item. There is no such thing as a perfect company or process, therefore the closest to perfect you can become is to recognize that it is a continuous process of improvement.

Furthermore, your organizational change agents do not want to foster false expectations that lean transformations are not difficult or that the effort is not long lasting. Again, this would be sending the wrong message, which can lead to high levels of unmet expectations, with a consequence that your workers lose their energy and motivation to make changes and improve. Also, it will not reinforce the notion that it takes discipline to implement lean transformations and sustain them over time. However, while there should never be a completion date for process improvements and the level of effort required should not be sugar coated, it is important for you to celebrate successes along the way as a way to mark progress and sustain momentum (Kotter, 1998).

To ensure that there is a clear understanding of what lean philosophy entails, you must send a clear message that learning by doing and making improvements is a cyclical and ongoing process that is modeled by management at all levels of your organization. In addition, you should not include completion dates on process improvement efforts. Instead, have ongoing recognition and celebrations of successes, and always include a section on next steps when reporting on improvement efforts, even if it means taking the learning that occurred in one work area and applying it in other work areas.

Challenge V: Taking on Too Many Initiatives at Once. A major problem in an organization's lean transformation is taking on too many initiatives at once. While your improvement team may be enthusiastic about potential improvements as they are discovering root causes of problems and brainstorming solutions, it is important for your team to recognize that slow, steady progress is more manageable and is the essence of incremental, continuous improvement. If an action plan is too complex and requires a considerable amount of effort and coordination among many players, it will increase the likelihood that someone will drop the ball, and that can derail an entire effort.

To make sure that too many initiatives are not untaken at once, your team leader or coordinator for improvement efforts must have the experience, knowledge, and skill set

Be not afraid of growing slowly; be afraid only of standing still.

— Chinese proverb

to (1) facilitate group processes, (2) motivate team members, (3) trouble-shoot when necessary, and (4) keep everyone on track with respect to making progress on the action plans.

Challenge VI: Being Clear on Process and Outcomes of Service Organizations. Service-related operations—whether those in the office or administrative functions within business and industry or social service organizations—have difficulty being crystal clear as to their processes and the end-product of their service. Unlike production systems, service operations (1) have information flows that are unstructured, (2) use informal work scheduling, (3) are not as likely to have standardized work processes, making their processes less stable and thus subject to a high degree of process variation, and (4) have workers who may support several value streams within your organization. Together, these characteristics of service operations make it less easy to map value streams and measure performance. Given these challenges, staff and stakeholders within the service sector might think that lean applies only to manufacturing where products are produced on an assembly line with more controlled processes.

To address this challenge of perspective, you must make sure that your staff and stakeholders recognize and accept that value streams and associated performance measures will not be perfect; instead, they will be approximations that are determined either through

1. observation of a process in action over a selected time and employee sample, making it a point-in-time set of descriptive data that are highly contingent on the particular time, place, and worker conditions, or through
2. educated guesses provided by those who implement a process about the steps and associated descriptive measures (e.g., cycle time, VA time, wait time, elapsed time, number of people involved, and resource cost).

Regardless, even in service environments, these approximations can serve as eye-openers with respect to the amount of waste embedded within processes. Therefore, in your lean transformation efforts, it is vitally important for you to emphasize the value of using lean mapping and analysis tools, as well as tools for gathering data to approximate the performance of your service operations.

As a word of caution, though: before your process improvement efforts are implemented or made part of the improvement process, processes must be stabilized. If not, the improvement results could be lost in the process variation that will occur. Finally, process improvements should be implemented on the low-hanging fruit as soon as possible to reinforce the positive results that can happen when lean solutions are applied.

Challenge VII: Selecting Improvement Projects Not Aligned with Organizational Goals. There are two challenges associated with the identification and implementation of improvement projects. First, when your process improvement efforts target processes that do not have real significance with respect to improving your organization's performance, there is a danger that the results will not produce significant gains in efficiency and effectiveness. When this happens, there can be ramifications with respect to the level of support and commitment from your upper-level management, worker enthusiasm, and motivation to make changes. It is important to select improvement efforts that are closely linked with your organization's service and delivery goals. Sometimes this disconnect can occur because your organization's goals are not clear; even worse, sometimes there may be conflicting goals at different levels or functions within the organization (Riley, 2008).

Second, if your improvement projects are selected within a functional area without concern for the impact that changes may have either upstream or downstream, there is a danger that improvements in one area can have a detrimental effect on other areas. Therefore, it is important not to lose sight of your whole organization and the performance measures you use to assess operational efficiency and effectiveness, because what you measure will drive behavior.

To prevent the selection of improvement projects that are not linked to organizational or business goals, projects should be ranked according to their relevance and impact on these goals and according to the improvement goals for a specific work area. If your organizational or business goals are not clear, then you must determine them prior to embarking on process improvement projects. Developing a logic model or developing a strategic plan are two approaches to use for clarifying organizational goals associated with different programs or functions within your organization; as discussed in Chapter 1.

Execution Challenges

Finally, there are three execution challenges that you must address for a successful lean transformation.

Challenge VIII: Inappropriate Improvement Team Membership. When you do not have the right composition for your improvement teams there is a danger that the scoping of a project will be too narrow, the ability to have an all-encompassing view of organizational goals will be limited, or improvement plans may be impractical due to resource constraints, lack of buy-in from frontline workers, or management opposition. The conundrum here is that an ideal size for an improvement team is six to ten members, because any fewer can result in the lack of representation from appropriate functional areas and any more can result in management difficulties and loss of focus (Riley, 2008). Moreover, you may have resource constraints or encounter roadblocks in work environments where the heavy demands on existing staff time make it difficult for your employees to stop what they are doing on a regular basis to make significant changes in work processes or to become involved in process improvement efforts.

To ensure that your improvement team has all the appropriate representation, the full-time team members can be limited to no more than ten members, and supporting team members can be brought in when needed. Also, minutes or documentation of team improvement activities should be distributed to all full-time and supporting team members to ensure they are kept in the loop as to your improvement effort. Furthermore, if your organization is truly committed to organizational improvement, you must bite the bullet and find the time and resources to engage in the process improvement efforts (e.g., by having a dedicated lean specialist), since doing so will ultimately free up the resources (time, money, and labor) that are trapped in a cycle of WAs.

Challenge IX: Ineffective Control Plans. When you do not have mechanisms in place to prevent returning to the way it has always been done, there is a high likelihood that worker behavior will revert to habit. If this happens, the benefits of the lean improvements are lost and your organizational members will have their opinions confirmed that lean transformation efforts are just another passing improvement program that can be tolerated until it goes away.

While you may establish more instructions, signoffs, and auditing procedures to control a new process, these will not be effective in the long run if your process improvement makes work more difficult and cumbersome. Careful consideration and ingenuity from your process improvement team, with close feedback and involvement of frontline workers, is a must to ensure that new processes (1) are easier and more enjoyable to run than the old, (2) do not have the associated URs that make work frustrating and undesirable, and (3) make going back to the old way of doing work difficult.

Challenge X: Inappropriate Management of Process Improvement Efforts. There are two challenges associated with the management of process improvement efforts. One is the human tendency to jump to solutions too quickly without taking the time to systematically follow the problem-solving process as described in this book. Instead of doing the mapping of processes, gathering of performance data, and determining root causes, your teams can become enthusiastic about what they see to be obvious solutions to process or performance issues. While some of their initial thoughts on ways to solve problems may be on target, it is important to proceed through the rational process of determining the current state, finding root causes of problems, and implementing trial runs of improvements to assess results.

Correspondingly, because doing follow-up measures of performance and using that data to make informed decisions about improvement efforts requires a disciplined effort, your teams may fall short of completing the full cycle of planning, implementing, evaluating, and revising based on results. When this happens, the full benefits of organizational learning and continuous quality improvement are lost.

To ensure that your process improvement teams engage in their effort systematically and complete the full cycle for their particular effort, it is important for your organization to have the right person(s) serve as the team leader and other coordinators at the work area(s) where improvements are being implemented. These persons must have the combination of experience, knowledge, and skill and be the doers, as described earlier in this chapter (individuals who are instrumental in making change happen). Specifically, the right persons should have the "experience in the business or industry, knowledge of the tools and techniques of lean thinking, and the soft skills that allow them to put that experience and knowledge to work" (Howardell, n.d., p. 1).

CRITICAL FACTORS IN CREATING A LEAN TRANSFORMATION

It is fruitless to say there is one factor that can explain the success and failure of a lean transformation. In fact, lean transformations are hard work; it is important to identify, first hand, how organizations experience their transformation as they move from "waste ridden non-value added processes and system to a lean organization" (Hagood, 2009, p. 1). With this in mind, the list of truths provided in Table 8.1 poignantly describes these experiences. Although many of these truths have been touched on in our discussion of the challenges and pitfalls, they bear being repeated. It is important as you embark on your lean journey to fully understand the minefield you are walking through. Having a thorough understanding of the good, the bad, and the ugly, as well as the words of wisdom embedded in these truths, will help you prepare for your journey and arrive safe and sound because of your preparation and forethought about the obstacles you might encounter.

Taking these truths as a starting point, and considering our discussion of the challenges of lean transformations, there are three key areas where you must focus attention to increase your likelihood of success in undergoing and sustaining a lean transformation within your organization. These three areas—**leadership**, **employees**, and **management system**—work synergistically to create the conditions under which success

TABLE 8.1. The 12 1/2 Truths of a Lean Transformation

1. Lean isn't easy to do ... if so everyone would have already done it!

If you anticipate that a Lean transformation is going to be easy, then think again. Many of the concepts are meant to be easy to grasp and most times they are, but it takes dedication, diligence, and hard work to make the transformation a reality over time. Many organizations that started out strong didn't have the stomach to see the transformation through. Why, because it requires sustained, consistent effort!

2. Waste must be viewed as the enemy #1 if you are to be successful.

Waste ... whether it be waiting, rework/defects, transporting, extra processing, overproduction, inventory, or motion, has to be viewed as the enemy. Target it for elimination! Tolerance of waste must become inexcusable in all processes in order to be successful.

3. Ownership of the process is a MUST at all levels of the organization.

A Lean transformation will fail if not embraced at all levels. Be aware that many times significant Lean ownership issues will surface even at the top levels of the organization. This can be the most dangerous given their subtleties which can undermine the efforts of those below them in the organization.

4. Ultimately there can't be an option to not embrace the philosophy.

You should take adequate time to bring everyone along in the organization with the Lean approach to doing business. This will take time but ultimately leadership must make it clear that eliminating waste and improving the operation is not optional.

5. Lean will succeed or fail based upon the organization's leadership.

I've always believed that everything rises and falls based upon leadership. Nothing has highlighted this more than observing Lean transformations.

6. Lean will highlight the strengths and weaknesses of leadership.

Do you have members of your leadership that are weak, can't make things happen, don't follow through in execution, don't communicate, or can't accept change? This will become more evident than ever during a Lean transformation. The good news is that strong leaders and managers will come to the surface more than ever before. Promote and highlight the strong!

7. Leadership's commitment will be tested early and often.

Leadership must walk the talk. Only paying lip service to the philosophy of Lean will be discovered quickly by skeptics and waste loving curmudgeons. Leadership will be watched closely so look for opportunities to make believers out of those on the fence. This does not mean everything must be warm and fuzzy, but it does mean you have to stay true to the Lean philosophy and purpose.

8. You will make mistakes! If made trying, then just try again!

The worst mistake you can make is to not try new systems and methodologies because you fear you'll make mistake. You have to make sure that your actions do nothing to negatively affect safety, quality of products, patient care/safety, but many times we are fearful of making mistakes that don't pose that danger.

9. Ongoing and honest communication is both critical and a must.

You can't communicate enough throughout the process, especially in the early days of your transformation. Your communication must be honest, painting a vision of what you're trying to accomplish. Once you think you've communicated enough, you're probably just beginning to get the vision through to the team.

10. Lean has to become more than a program or a few events. It must become a way of life, which permeates all levels.

Many organizations do a few "fly by" Kaizen, Value Stream, Rapid Improvement, or 5-S events in selected areas and consider the job done. All that will do is get the organization initially excited only to be let down from lack of sustainability. That is not Lean thinking. These are elements and tools of Lean that must be used but for the transformation to be successful an organization must make eliminating waste in all forms and sizes a part of its culture and not just part of isolated events.

11. Lean isn't cheap or a quick fix.

Many organizations think they can shortcut the process by cutting corners. Others think Lean is a quick fix and cheap. Both thoughts are incorrect. Lean will lower your costs and improve quality and customer service over the long haul but it will not be quick or cheap.

12. Lack of timely action or follow-through will cause the process to fail.

Nothing will kill buy-in and commitment from the front line troops faster than leadership and management not following through on its commitments to the transformation, including any "work-outs," events, action items, etc. ... It's appropriate for management to say no to some ideas, but to not follow through or stall the process, due to lack of attention and commitment, will be the death knell of your transformation and destroy critical buy-in at all levels of the organization.

12 1/2. Most Important! And No Half Truth! Lean is ongoing and never ending!

The process of improving never ends. A Lean transformation has no end date! The process is ongoing and is never a closed out action item. There is no such thing as the perfect company or process, therefore the closest to perfect you can become is to recognize that it is a continuous process of improvement.

can be achieved, although they may vary with respect to their level of importance, depending on the situation. Given the varying levels of importance, our discussion of these areas and the steps that are necessary to transform your organization are not presented in terms of one being more important than another.

Role of Leadership

All organizations need good leadership, but the greatest need for effective leadership is in times of change, particularly when an organization is attempting to transform itself. In light of this, we first examine qualities or traits of effective change leaders. Next we describe different levels of leadership and the contributions that leaders make at each of these levels to sustain a lean transformation.

Qualities of Effective Leaders. Great leaders are not only those at the top of an organizational hierarchy—they can exist at all levels within an organization. While some leaders may have visions that are more far-reaching than others or that have the responsibility to motivate more people, there are some basic characteristics they possess. Regardless of their position within your organization's hierarchy, effective change leaders have these traits (Kotter, 1998, p. 3):

1. They are able to see situations through a different lens than others and often challenge the status quo.
2. They are energetic and able to handle challenges easily.
3. They possess a quest for learning and are driven by goals and ideals—the engine that pushes them toward continuous learning.
4. They continue to take risks, going beyond their own comfort zones, even after they have achieved a great deal.
5. They are open to people and ideas, even at a time when they might think, because of their successes, that they know everything.
6. They are deeply interested in a cause or discipline related to their professional arena and are motivated by deep emotions, not just by intellect.
7. They tap into deep convictions of others and connect those feelings to the purpose of the organization, i.e., they show the meaning of people's everyday work to that larger purpose.
8. They invest tremendous talent, energy, and caring into change efforts.

Leadership Roles in Sustaining Lean. While leadership capabilities can exist at all levels within your organization, individuals within senior- and middle-management leadership positions have complementary and overlapping roles they must play for a lean transformation to be successful. Figure 8.2 shows leaders at three organizational levels, along with their primary contributions and associated tasks with respect to sustaining lean (Mann, 2009). It is essential that leadership within your organization, at the strategic, programmatic, and tactical levels, do the following:

1. **Establish a strong guiding coalition.** Given the nature of today's complex organizations, where hierarchical, command-and-control management practices are less common or useful, your leaders must win the support of employees and other stakeholders for change initiatives. If you do not establish a strong guiding coalition with all the requisite representation, unexpected resistance can surface and derail your change effort. Three strategies for creating your strong guiding coalition are (1) engaging the right talent, (2) growing the coalition strategically, based on representation from important stakeholder groups, and (3) working as

> *All true leaders have learned to say no to the good in order to say yes to the best.*
>
> — John Maxwell

> *Leadership and learning are indispensable to each other.*
>
> — John F. Kennedy

FIGURE 8.2. Organizational Roles and Contributions to Sustain a Lean Initiative

Leadership Roles in Sustaining Lean				
Organization Level	Primary Contribution	Tasks	Secondary Contribution	Tasks
Strategic: Senior (CEO, Sr. VPs)	Governance; Steering and oversight	Support for a crossboundary perspective	Measurement; Adherence to post-project processes	Monitor intersection measures; *Gemba walks*
Programmatic: Function (VPs, Directors)	Accountability	Meet project commitments; Manage intersection performance	Disciplined adherence; Commitments to processes post-project	Collaborate in process management; *Gemba walks*
Tactical: Department (Managers, Supervisors)	Tactical Lean Management System	Disciplined adherence; *Gemba walks*	Associate engagement; Continuous improvement	Teach, practice root cause problem solving

Source: From "The Missing Link: Lean Leadership" by D. Mann, 2009, *Frontiers of Health Services Management, 26*(1), p. 16. Copyright 2009 by Health Administration Press. Reprinted with permission.

> *Never lose sight of the fact that the most important yardstick of your success will be how you treat other people.*
>
> — Barbara Bush

a team, not just as a collection of individuals. The right talent means having individuals who have the necessary skills, experience, and chemistry to work together. Just because someone is in charge of a department, division, or other stakeholder group, does not mean that he or she is the right person for the coalition. It is critical to identify coalition members that "have a strong position of power, broad experiences, high credibility, and real leadership skill" (Kotter, 1998, p. 3).

2. **Address cross-boundary difficulties.** Difficulties may arise when process improvements cross over functional divisions within an organization. Therefore, it is essential that your senior leadership establish governance mechanisms that cross divisional boundaries and build support for your organization-wide vision; that vision must focus on designing value-producing processes that meet your clients' needs. Such a vision is an essential component of any change effort, therefore you must create a vision, as well as develop strategies to achieve the vision. However, defining a vision is not always a rational process and is often more emotional than rational, which requires tolerance for different views and the ability to reach consensus. Regardless, having a vision provides a basis on which to resolve conflicts and tensions—it is the beacon that guides the decisions about actions to take and helps to prevent organizations from getting bogged down in "I win, you lose" fights. It is important for this vision to be clear in terms of intention, appealing to stakeholders, and ambitious yet attainable. While your vision must be focused enough to guide decision making, it also must be flexible enough to accommodate changing conditions and individual initiative.

3. **Support and sustain new behaviors and practices.** It is the responsibility of your leadership to hold everyone accountable, to monitor progress through regular reporting of process-focused measures, and to recognize improvements and successes of your improvement team efforts. However, while it is important to celebrate successes in organizational transformations, it is just as important not

to mislead anyone into thinking that success is just around the corner or that the change process will not be difficult. If your leaders send a misleading message about either the duration or level of difficulty of the change process, the results can be detrimental to the development of a system of sustainable success.

4. **Create conditions that reinforce a culture of continuous improvement.** Your leadership must reinforce the notion that continuous improvement is the norm rather than the exception, though a disciplined adherence to improvement efforts and *gemba walks*. (Gemba walks are regular walks managers take through your organization to make observations and gather input from the frontline workers regarding process improvement opportunities.) Your leaders will need to establish a sense of urgency for change efforts, at the same time avoiding declaring premature victory once an initial goal is met, which can dampen a commitment to continuous improvement.

In line with this overview of the role of leadership in a lean transformation, Kotter (1995) identifies eight steps to transform your organization. These eight steps are shown in Figure 8.3; each of these steps reinforces our discussion of the role of leadership in a lean transformation.

Role of Employees

According to Edwards Deming, the father of the modern quality movement, most people are willing workers who are subjected to working in processes that have waste built in. Without calling it waste, most workers accept their WAs as a normal routine. Therefore, to engage employees in a concerted effort to eliminate waste, you must establish an organization-wide mindset that admits that all processes contain waste, put tools in place to identify waste in a nonblaming environment, and allow employees to eliminate it.

While this sounds easy, as previously discussed there are employee challenges you must face eventually to establish the mindset within your organization to rigorously attack waste in your processes. Regardless, employees play a significant role in the change process and you must take steps to eliminate or reduce the barriers that prevent employees from being full partners in your organization's lean transformation. In essence, you need lean people to create a lean enterprise (Howardell, n.d.).

Engaging Employees in Lean Transformations. As previously described, Tapping and Dunn (2006) identify different types of employees based on their level of support for change efforts: the doers, helpers, followers, resisters, and saboteurs. To ensure that employee problems do not derail your lean transformation, the focus should be on converting a sizable portion of the resisters and saboteurs. The issue is how to make that conversion. As already discussed, it is critical that your employees be engaged at the beginning of a change effort. They must play a primary role in identifying what is important to them, as well as be involved in problem-solving activities. In addition, employees must be made aware that a lean transformation will require change (for the better). Essentially, the changes must be implemented in such a way that your employees buy into the transformation process and see it as a way to reduce the hassle factors in their work.

It is not unusual for employees to be unaware that change needs to occur. For example, Figure 8.4 identifies a number of prevailing thoughts and opinions among employees that would limit their acceptance of the need for change and their willingness to participate in change efforts (Tapping & Dunn, 2006). When these thoughts

> *Your expression is the most important thing you can wear.*
>
> — Sid Ascher

FIGURE 8.3. Kotter's Eight Steps to Transforming Your Organization

Eight Steps to Transforming Your Organization	
Establishing a Sense of Urgency Examining market and competitive realities Identifying and discussing crises, potential crises, and major opportunities	**1**
Forming a Powerful Guiding Coalition Assembling a group with enough power to lead the change effort Encouraging the group to work together as a team	**2**
Creating a Vision Creating a vision to help direct a change effort Developing strategies for achieving that vision	**3**
Communicating the Vision Using every vehicle possible to communicate the new vision and strategies Teaching new behaviors by example of the guiding coalition	**4**
Empowering Others to Act on the Vision Getting rid of obstacles to change Changing systems or structures that seriously undermine the vision Encouraging risk taking and nontraditional ideas, activities, and actions	**5**
Planning for and Creating Short-Term Wins Planning for visible performance improvements Creating those improvements Recognizing and rewarding employees involved in the improvements	**6**
Consolidating Improvements and Producing Still More Change Using increased credibility to change systems, structures, and policies that don't fit the vision Hiring, promoting, and developing employees who can implement this vision Reinvigorating the process with new projects, themes, and change agents	**7**
Institutionalizing New Approaches Articulating the connections between the new behaviors and corporate success Developing the means to ensure leadership development and succession	**8**

Source: From "Leading Change: Why Transformation Efforts Fail," by J. P. Kotter, 1995, *Harvard Business Review,* *73*(2), p. 61. Copyright 2000 by Harvard Business School Publishing. Reprinted with permission.

exist, your employees will perceive the status quo to be good enough, and will be hesitant to participate in teams, particularly if, in the past, the teams had limited success. If these conditions exist within your organization, it is vital to have an open and honest dialogue to address the questions your employees have. If the issues are not addressed early on, they can inhibit team unity for kaizen events and jeopardize the long-term success of your organization's lean transformation.

FIGURE 8.4. Employee Thoughts About Change Efforts

Seeing the Challenge

Many employees may not realize that a change needs to occur. Many may think the following:

Skills Employees Must Have. In addition to these recommendations regarding how to engage employees in the lean transformation, your employees must have a number of skills if your organization is to fully embody lean thinking and walk the talk. As a prerequisite to effectively apply lean thinking and tools within your organization, it is vital that your employees have the following skills (Howardell, n.d., pp. 3–7):

1. **Client consciousness.** Lean thinking starts with specifying value from your client's perspective. To do this, employees must know who your client is and what your client wants and expects. In addition, the idea of an internal client must be on the radar screen, because in some jobs employees only provide

Also, these types of thoughts may be prevailing:

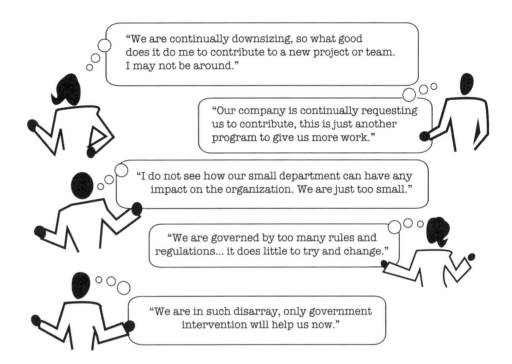

These are very pessimistic statements, but they are true in many instances. Regardless of the type of industry, where the organization is in its life cycle, or what external influences may play a role, there are numerous reasons why the status quo is perceived to be good enough. However, employees must be made aware that a lean office will require change (for the better). Employees may have previously been part of teams that had little success in their implementation. An honest dialogue must occur to address the many questions that linger in employees' minds. Concerns must be addressed early on in any lean project to promote team unity.

Source: From "Lean Office Demystified," by D. Tapping and A. Dunn, 2006. Chelsea, MI: MCS Media, Inc. Copyright 2006 by MCS Media, Inc. Adapted with permission.

direct support for others within your organization, and they do not have direct interface with your external client. Therefore, anyone who is the direct recipient of the output from a process should be viewed as a client. Once employees know who their clients are, they must implement the means to know what the clients want and need, as well as measure the extent to which these wants and needs are being met. In a lean organization, employees keep in touch with their clients, identify barriers to client satisfaction, and eliminate these barriers.

2. **Organization-wide (enterprise) thinking.** In thinking about your organization as a whole, your employees must shift away from functional or departmental thinking. When employees only think within their department or functional

area, they are limiting their ability to perceive how change will impact the whole, rather than them personally, within their own department. If there are silos within which your employees exist, it increases the chance for turf battles wherein employees blame others for problems, spend hours to document why a problem is not their fault, and find ways to make sure changes do not affect them. For your employees to grasp enterprise thinking, it is essential that they understand that all work is accomplished by a process that can transcend several functional areas to convert inputs into outputs and deliver value to your clients. Hence, if there is a problem, the task is to examine the process to determine where the waste occurred, and not **to determine the names of the individuals involved**. To ensure enterprise thinking, it is essential that your management, as well as the line workers, have a basic understanding of the concepts and methods of process improvement, process mapping, performance measures, and process redesign.

3. **Adaptation.** Being adaptive is one of the most critical skills for people within an ever changing environment, which will accelerate once your organization embarks on its lean journey. Lean people need to know how to adapt a changing environment to their advantage. For example, if client demands are constantly shifting, your organization must be agile and shift your service and products to meet the new client needs. Individuals can usually be placed on a continuum of reaction to change from one of resistance to one of positive acceptance, as discussed earlier. If a person is on the resistance end of the continuum, then steps must be taken to move that person along the continuum so that he or she becomes more adaptive to change.

4. **Taking initiative.** The heart of a lean organization is that your employees have internalized the norm of personal responsibility when it comes to identifying the WAs and the URs of their efforts. Furthermore, they bring those issues forward to discuss root causes and options for eliminating waste, and to take action to resolve the problems. In essence, this is problem-solving at the grassroots level, rather than having only high-level teams study the problem, come up with solutions, and impose those solutions on others. To be capable of taking initiative, your workers must have the foundation of knowledge about what it means to be a lean organization and the tools that are available to engage in the process of problem resolution. The knowledge and skill set must include the entire cycle of planning, implementing, evaluating, and taking corrective action. Furthermore, lean people take the initiative to "maximize their productivity, manage their time, and stay organized . . . at the personal level as well the enterprise level" (Howardell, n.d., p. 6).

5. **Innovation.** In the lean approach to problem solving, your employees will be called on to invent new processes, and perhaps new services and products, as required by client needs and wants. For this to occur, creativity is a must across the organization, not just a skill of those in management or research and development divisions. Thus, as a lean organization, you must take the initiative to foster creative thinking among your employees. Again, you can accomplish this by educating them on the lean concepts and methods, as well as by establishing an organizational structure that nurtures experimentation and learning by doing. At a minimum, the knowledge and experience that your employees must have

Failure is an event, never a person.

— William Brown

146

The healthiest competition occurs when average people win by putting above-average effort.

— Colin Powell

include how to (1) to analyze problems, (2) apply critical thinking skills, (3) apply analysis techniques such as value stream and process mapping, 5-Whys, Fishbone Diagrams, and charting or graphing performance data, (4) engage in brainstorming and creative thinking techniques, (5) overcome barriers to creative thinking, and (6) recognize different thinking styles and when to apply them.

6. **Collaboration.** If your organization is to be agile and capable of responding to problems or issues quickly you must have collaboration between individuals and groups; that collaboration is an important component of your lean organization's strategy. Having a cumbersome hierarchical structure and chain of command can stifle creativity and action at the grassroots level, which is at the heart of a lean organization. In terms of ensuring that collaboration is a reality, there are a number of actions that your management must take to define the expectations and limits on the collaborative groups: (1) determining why they are creating the groups, what groups they are creating, and if these groups are cross-functional or departmental based; and (2) deciding what authority the groups have, how the groups will be measured and rewarded, and how individual performers will be recognized. Once your management has established these guidelines, the collaborative group members must be trained on (1) stages of group development, including storming, forming, norming, and performing (Tuckman, 1965); (2) group roles of leader, scribe, and process observer; and (3) consensus decision-making strategies.

7. **Influence.** As a lean organization, you need to have everyone within your organization focusing on your goals and taking action to ensure these goals are achieved. This will require your leaders to make tactical operational decisions that are aligned with your organizational goals. However, your leaders may not be in formal leadership positions; they may be individuals that have influence over others, but without formal leadership responsibilities. Therefore, you must identify these leaders, ensure they are supportive of your lean transformation, and secure their commitment to use their influence to move the organization in that direction. These influencers must have the following understanding and skill set: (1) know what it means to lead, (2) know how to take leadership actions, (3) know how to create and share a coordinated vision, (4) know how to align the organization on what needs to be done, and (5) know how to empower people to get things done.

Like lean tools discussed in earlier chapters, these skills are not new in themselves, but a number of them are given only lip service within organizations. To truly be a lean organization, these skills must be taken seriously. According to Howardell (n.d., p. 3), these "lean people skills should be viewed as a set. Everyone working in a lean enterprise requires them all. A weakness in any skill is the proverbial weak link. It is a flaw that must be corrected." To fully achieve the designation as a lean enterprise, your organization must implement strategies to develop the skill set in its people. The plan starts with training, followed by more training and opportunities for the people to practice what they learn. It is imperative for your organization to set aside resources (time and money) to do this. An excellent example of how one community addressed a need to train its public and social service sectors in process improvement methodologies (both lean and Six Sigma) is described in Example 8.2.

EXAMPLE 8.2. **Building Community Capacity for Process Improvement: Buffalo, New York, Experience**

There are a number of ways that individuals and organizations are able to address their educational needs with respect to building their capacity for process improvement. The proficiency in process improvement methodologies may be achieved through formal training, learning by doing, self-education, and consulting services.[1] The example presented herein showcases two models that coexist in the Buffalo, New York, community. These two models demonstrate community-wide capacity building initiatives that have a wide reach throughout county government and nonprofit social service sectors, exemplify the benefits of collaboration among community stakeholders, and vary with respect to the employed approach.

University at Buffalo Center for Industrial Effectiveness, Canisius College, and Erie County Government
Beginning in 2008, Erie County, New York, instituted a new division, Six Sigma Implementation. The goal of this division was to reform government operations through application of Lean Six Sigma methodology. The director of this new division, Alfred Hammonds, Jr., was a senior project director at the University at Buffalo's Center for Industrial Effectiveness. With $120,000 of grant funding for the first year of the initiative, Hammonds was able to train nine green belt–level employees and eighteen Six Sigma champions comprising commissioners and department heads (Spina, 2008).

To support the educational needs of Erie County government, two local colleges—University at Buffalo Center for Industrial Effectiveness and Canisius College—provided needed expertise and educational programs. Both educational institutions responded with the development and deployment of the Lean Six Sigma Green Belt Certification programs and other program offerings that supported wide organizational change. The teams of faculty also were readily available to facilitate on-site implementation and learning on an as-needed bases.

In 2008, Hammonds estimated $900,000 savings from the green belt process improvement projects that were initiated and completed during the first year (Spina, 2008). The total spending of Erie County government for Lean Six Sigma initiative in 2008–2009 was $711,000. The funds were utilized to support the position of the director of the initiative, staff training, software, and project support. The savings from the green belt projects during the same period totaled approximately $7.7, million equating to a return on investment of almost eleven-fold (American Society for Quality Buffalo, n.d.).

The mutually beneficial relationship between Erie County and both educational institutions continues to exist and the benefits of the streamlined government operations are enjoyed by the residents of Erie County, New York. The green belt projects for 2010 include a wide variety of county services such as mental health, parks, golf courses, social services, public works and highways, sewers, health, and county correctional facilities and jails (American Society for Quality Buffalo, n.d.).

United Way and Nonprofit Organizations
At the same time, recognizing the strategic importance of Six Sigma training and drawing on the available community resources and goodwill, the United Way of Buffalo and Erie County designed its own educational model to address the lack of knowledge and skills in the area of process improvement within the nonprofit sector.

Recognizing the financial limitations of nonprofit organizations, the United Way offered free Six Sigma yellow belt training, utilizing the volunteer service of a retired Honeywell executive, William J. Hill, PhD. Dr. Hill, a master black belt, was formerly a director of Six Sigma at Honeywell and a director for the Center for Quality and Productivity Improvement at the University of Wisconsin. Together with Joe Roccisano of the United Way, Dr. Hill has trained five cohorts of individuals from twenty-four nonprofit organizations. To date, these cohorts have worked on thirty process improvement projects. The organizations participating in the training include Consumer Credit Counseling Service, Benedict House, Catholic Charities, Community Services for the Developmentally Disabled, Food Bank of Western New York, Meals on Wheels for Western New York, United Way of Buffalo & Erie County, Buffalo Municipal Housing Authority, Buffalo Urban League, Compass House, Gateway-Longview, Hispanics United of Buffalo, Martin House Restoration Corporation, New Directions Youth and Family Services, Albright Knox Art Gallery, Baker Victory Services, Catholic Charities, Mid-Erie Counseling & Treatment Services, Spectrum Human Services, American Education Foundation, Community Concern of Western New York, Family Justice Center, Joan A. Male Family Support Center, King Urban Life Center, and Upstate New York Transplant Services (United Way of Buffalo & Erie County, n.d.).

During the training, participating nonprofit organizations select a process improvement project within their organization and establish a cross-functional project team with up to five team members. This team from each nonprofit organization attends a two-hour introductory session, followed by four classes, each four hours in length, approximately two weeks apart. The overall class time totals eighteen hours. Teams work on their projects during and between classes.

---

[1] Lean and Six Sigma process improvement methodologies are often coupled in certification programs where individuals can earn a yellow belt, green belt, black belt, or master black belt. Together, they are methodologies that overlap, since both focus on eliminating inefficiencies in a work environment. Using one methodology or another is dependent on the process issue.

Some results nonprofit organizations achieved through the United Way's Six Sigma training include the following (United Way of Buffalo & Erie County, n.d.):

- Reduction of no-shows by 20 percent, with productivity increase of 30 percent
- Reduction of defects by 95 percent, with costs reduction of $19,000
- Two-fold improvement in compliance
- Two-fold increase in number of people served
- Increase in number or clients qualified for assistance by 64 percent
- Reduction of cycle time leading to revenue increase of $125,000

The Six Sigma yellow belt training offered by United Way is a true community asset that provides ongoing training and project support to nonprofit organizations of Erie County, New York.

It is our choices . . .
that show what we
truly are, far more
than our abilities.

— J. K. Rowling

Role of the Management System

All organizations are dynamic and evolve over time as a result of reactions to an external environment, and the internal interrelationships between and among an organization's functional units and the people that constitute those units. Obviously, it is in the best interest of your organization to plan for change and to not be caught off guard when you experience problems or challenges to your existence. The management system within your organization is the mechanism through which planning occurs. Moreover, it handles the dynamics between organizational leaders or managers and employees. The discussion below reinforces many of the points we have already made with respect to the role of leaders and employees in lean transformations.

Planning Models. Traditional planning methods focus a great deal of attention on the development of strategic plans, where selected organizational members and stakeholders spend most of their time answering questions similar to those of a "SWOT" analysis (Armstrong, 1982). In this strategic planning approach, the focus is on identifying your organization's **S**trengths, **W**eaknesses, **O**pportunities, and **T**hreats. In addition, your organization identifies its values and social responsibilities. Then, based on the answers to these questions, your planning group makes decisions about strategies to achieve your stated mission or goals.

Too often, that is where the effort stops. Your strategic plan is done, it is printed and distributed, but there is no action plan to ensure the strategies are implemented. In an effort to fill that gap in lean change efforts, Dennis (2006) presents his model of strategy deployment. His model embodies the fundamental tenets of lean thinking with respect to the roles of the leaders and employees and how problems are resolved and improvements made. Figure 8.5 offers a pictorial representation of the conventional versus the lean model of planning, adapted from Dennis's model, which was developed for a manufacturing environment.

An organization that is based on a command-and-control approach to management is in total opposition to what lean philosophy teaches us about the importance of worker empowerment (Mitchell, 2000). Command-and-control management approaches can suck the willingness out of employees to do their best and buy into their organization's goals and objectives. Command-and-control is entrenched in many organizational structures and lives in the minds of supervisors, managers, and executives that it is their responsibility to tell people what to do, to keep information about operations to themselves, and to stifle individual ingenuity and creative problem solving. Because of this history, getting employee buy-in is one of lean's biggest challenges, as previously discussed.

FIGURE 8.5. Conventional versus Lean Mental Models

Conventional Planning Mental Models....

1. Leader uses command and control to manage workers.

2. Leader stays isolated from front-line workers and knows little about how work is done.

3. System has some standards, but workers are unfamiliar with them and compliance with standards is unknown.

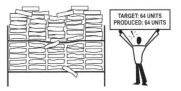

4. Focus is on outputs and meeting the targets for quantity, regardless of quality; services and/or products delivered have errors/mistakes.

5. People point fingers at others for mistakes.

6. Specialists are invited to solve problems and they do it by using complex methods.

...Going nowhere fast.

Without a doubt, organizational change can and does happen regardless of the management philosophy underlying it. But instead of using a traditional command-and-control approach to manage your lean transformation—an approach that does not embody the underlying tenets of lean with respect to the dynamics and relationships between an organization's leaders, managers, and employees—your system of planning, execution, and management should have these features, as exemplified in Dennis's (2006, p. x) strategy deployment system:

LEAN Mental Model

1. Leader seeks input from workers and is a teacher and coach.

2. Leader routinely observes processes, asks questions, and gathers input from front-line workers and other stakeholders.

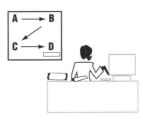

3. System has clearly defined processes that are simple and visually represented for important activities; it is obvious when a standard is not met.

4. Processes stop when quality is jeopardized to determine reason and make corrections; problems are not passed to the next in line or to the client/customer.

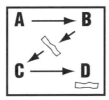

5. Problems are made visible and it is the process that has problems, not the individuals involved in the process.

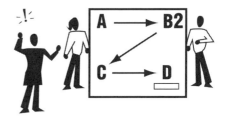

6. Everyone solves problems using simple methods.

...getting the right things done.

Source: From "Getting the Right Things Done: A Leader's Guide to Planning and Execution," by P. Dennis, 2006, Cambridge, MA: The Lean Enterprise Institute, Inc. Copyright 2006 by the Lean Enterprise Institute, Inc. Adapted with permission.

strategy deployment will focus and align your activities, and allow you to respond quickly to threats and opportunities. Moreover, it's a *human* system. People respond because it acknowledges their individuality. With strategy deployment, we don't *tell*—we *ask questions*. We don't *command*—we *engage*. Our people are not human *resources*—they are *human* resources. Most important, we try not to bore with meaningless data—we try to tell interesting *stories*. (emphasis in original)

Management Systems for Establishing Continuous Quality Improvement as the Norm. In addition to having a management system that increases buy-in from all stakeholders involved, you must have a way to achieve the positive results that you want (i.e., sustainability of lean improvements, leadership at all levels, employee ownership of their processes). The fundamental question you need to ask is, How do you manage process performance and allow the owners of processes to continuously improve them? At a basic level, you need to identify ways to

1. invest in developing people to be good problem solvers and to have at their hands the knowledge and tools to solve problems, e.g., the measures of performance and the target goals to achieve;

2. have a standard management process that tracks performance and shows any gap between current performance and expectations (i.e., goals and targets); and

3. have an expectation that the bar will be continuously raised (i.e., the expectation of continuous quality improvement, which means moving toward perfection and excellence).

In addition to the management systems that you can use to increase employee buy-in and understanding of what it means to engage in continuous quality improvement, there are techniques you can use to ensure that the workers have a true sense that they own their process, can make improvements in the way they do their work, and are able to sustain the improvements that are made. Having this sense is an essential part of sustaining lean within an organization. Specifically, your organization should do the following (Butler & Snyder, 2001):

1. Maintain a high level of attention to detail, as required, when your organization is undergoing a lean transformation.

2. Implement behavior-based strategies that have a built-in consequence-driven support structure that will reinforce the right behaviors associated with a new process until it becomes integrated into your organization's culture.

3. Use a coordinator in each work area to act as a team leader and to be part of a steering team of other coordinators and management representatives.

4. Train these coordinators and all management within your organization in the use of behavior-based strategies of observing behavior and applying methods of positive reinforcement until the employees' behaviors have reached habit strength.

5. Establish work area teams who are responsible for setting your team goals, pinpointing behaviors to observe, planning and implementing the celebrations when goals are reached, and determining new goals when the team is ready to move into new areas of improvement.

6. Ensure that your coordinators and managers implement the behavior-based strategies themselves, with the help of a facilitator as necessary, until their skill level in the use of these reinforcement strategies has been established.

7. Begin this process of employee-owned and employee-driven behavior-based strategies at a very simple level (e.g., having all employees implement 5S in their own work area), as they are learning how to communicate, how to reinforce, and how to get comfortable talking with fellow workers about their behavior.

8. Have management support this system by recognizing when workers are doing a good job, by attending meetings of steering teams, and by giving employees enough time to do the required observations.

> *The wicked leader is he whom the people despise. The good leader is he whom the people revere. The great leader is he of whom the people say, "We did it ourselves."*
>
> — Lao-tzu

152

9. Have management show support by saying the right things, at the right time, and at the right frequency, particularly as these things relate to giving work teams the authority to make the observations and take corrective actions, allocating the resources they may need to implement improvements and sustain them, and granting the time they need to engage in this oversight and problem-solving behavior.

Assessing the Lean Management System. In line with the basic tenets of lean thinking, all processes should be analyzed to assess the extent to which they are both effective and efficient in achieving their goals. If the processes in place are not performing as expected, then adjustments must be made to improve the performance. It is important to maintain the continuous change cycle of planning, implementing, evaluating, and taking action to make improvements. It is essential for your organization to apply this same thinking to the lean management system you establish.

The benefits of assessing your lean management system are three-fold (Mann, 2005). First, the assessment rubric (i.e., the criteria used in the assessment and the specific questions asked) establishes a set of standards for the lean management system; these standards clearly articulate the features of the management system, how it operates, and the results expected. Having this knowledge will help your organization in the design and implementation of your system.

Second, the comparison of an established lean management system (actual) to a system that sets the standard (expected) can provide your organization with a gap analysis, pinpointing areas where the standards are met and not met. As assessments are made over time, your organization can see how much you are improving.

Third, the results of the assessment that show where your organization is not meeting specific standards can be used to determine the next steps and areas to target with respect to making improvements. Your organization may take incremental steps to improve your management system, depending on the extent to which the standards are met or not met.

There is nothing new to this rationale as to why it is important to conduct assessments and how these assessments can be used to improve organizational performance. The difference here is that we are assessing the system established to manage the continuous quality improvements processes within an organization. Fundamentally, as expressed by Mann (2005, p. 157), "Defining expectations and holding people accountable to them is the key to a successful lean implementation. The higher in the organization this extends, the better the chances for success."

As for the criteria that are important to include in such an assessment, Mann (2005) has developed an assessment tool that offers eight categories of process and behavior, along with five levels of the system's operational status. Together, these form the structure of the assessment tool. The specific categories and levels are (Mann, 2005, p. 165):

Categories in the lean management system assessment:
1. Leader standard work
2. Visual controls for production [service delivery]
3. Visual controls for production support [service delivery support]
4. Daily accountability process
5. Process definition

6. Disciplined adherence to process
7. Root cause problem solving
8. Process improvement

Levels in scoring the lean management assessment:
Level 1: Pre-implementation
Level 2: Beginning implementation
Level 3: First recognizable state
Level 4: System stabilizing
Level 5: Sustainable system

The details of this assessment system with a complete copy of the tool is available in the appendix of Mann's book (2005). It provides the basic criteria and approach for conducting assessments that can be used across industry sectors. These assessments must be done by making direct observations and asking questions in employee work areas. With respect to the timing of the assessments, it is important to "just do it" as soon as possible to start out, since this will give you a baseline that can be used as a point of comparison for the periodic assessments that are done subsequently. The timing on the follow-up assessments may depend on the particular situation, since you do not want the assessment process to be burdensome and overly bureaucratic. It must be a straightforward process that produces immediate feedback that can be used to identify the next set of improvements. Conducting the assessments every ninety days is recommended.

Your assessors should come from all levels of leadership within your organization, since this provides all of them with first-hand knowledge of the criteria and how those criteria will be judged. Rotating your assessors is recommended, again, so that no single person has this responsibility. To ensure your assessors are comparable across each other with respect to their inter-rater reliability, the assessors must be trained and must engage in paired practice to determine if they are similar in their ratings.

To interpret the results of the assessment, you should report the data for each of the eight categories: this will provide guidance on the areas that are not yet up to standard. Figure 8.6 offers a visual way to display the results using a radar screen profile format that shows the extent to which the different components of your lean management system have been implemented. In Figure 8.6, we see that the majority of categories in this assessment of lean management are at the "first recognizable state." Only two, process improvement and root cause problem solving are at the "system stabilizing" level; disciplined adherence to process is at the "beginning implementation" level.

In summary, sustaining improvements over time may require a shift in your organizational culture such that the new way of operating becomes normative. We know that this shift does not happen overnight, and that it requires the leadership, as well as the employees, to buy into the changes. Without the commitment of all involved, it is easy to revert to old habits. Therefore, it is important to have a management system that embraces the observation of work areas to ensure that the order, visual controls, and standardization are maintained, e.g., through gemba walking. However, you must recognize that no work environment will ever have 100 percent of NVA activities eliminated, although a reasonable goal would be to work toward a level where 85 percent of them are. Therefore, you must continuously examine work processes for areas where waste can be eliminated and improvements made to increase your operational efficiency, effectiveness, and overall performance in delivering services to your clients.

FIGURE 8.6. **Score Profile Radar Chart for Lean Management Assessment**

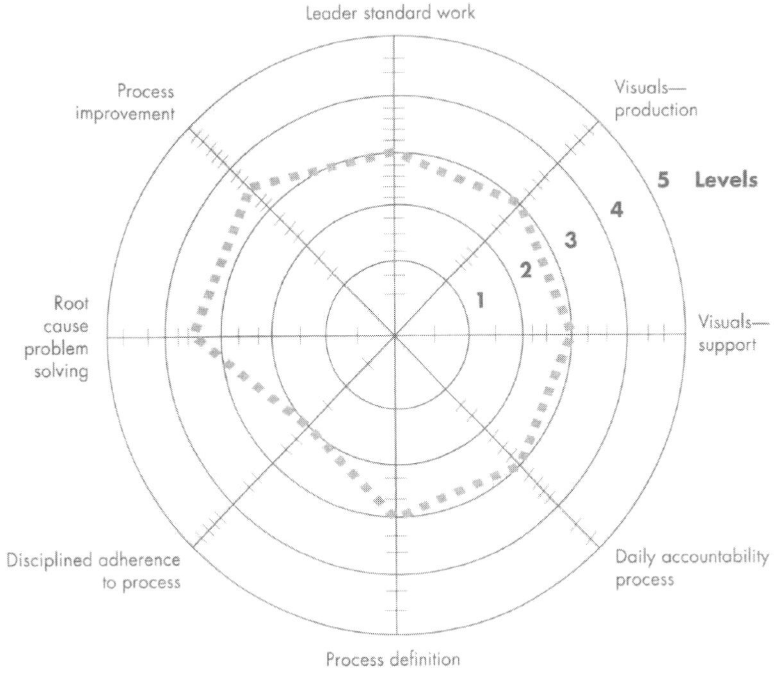

Source: From "Creating a Lean Culture: Tools to Sustain Lean Conversions," by D. W. Mann, 2005, New York: Productivity Press. Copyright 2005 by the Productivity Press. Reprinted with permission.

9

CHAPTER 9

Case Studies of Lean Transformations in Service Organizations

LEAN AND SERVICE ORGANIZATIONS: A RECAP

Our overview of lean concepts, methods, and tools as presented in this book provides only an introduction to how they may be applied in service organizations. Undoubtedly, there are many nuances and more to lean thinking than the concepts and methods presented in this book. Regardless, this discussion offers a glimpse as to how your service organization can improve its performance and benefit from the process improvements that are implemented as a result of lean thinking.

As discussed in earlier chapters, lean transformations occur over the long term; to be effective, this way of thinking must become embedded in your organization's culture. However, particular lean projects are generally event-based processes (i.e., a kaizen activity), best done over a period of three to five days, with the involvement of key players within the work process being mapped, along with a facilitator that has expertise in lean thinking and methods. In lean terminology, this is known as conducting a baseline event (Carreira, 2004).

The deliverables from a baseline event include the following (Carreira, 2004, p. 77):

1. A clear picture of the current state of a process (AS IS).
2. An equally clear picture of a future state (TO BE).
3. A specific definition of the waste in the targeted process.
4. A highly defined illustration of the disconnects across system processes.
5. A visual picture of improvement opportunities.
6. A prioritized plan of activities to implement the change.
7. Quantification of the expected results and the cost to implement (payback or return on investment).
8. An energized group with one collective mind and new purpose.

Once a baseline has been completed, the implementation stage of your lean transformation begins. Ideally, your leaders and management support the change effort, and a team within your organization spearheads the transformation process. However, you may need short-term technical assistance by a lean specialist to ensure that your effort stays on track and, when challenges or roadblocks are encountered, the lean specialist can assist in their resolution. The ultimate goal of this process is to build the capacity of your organization's employees to engage in lean thinking with all of their work processes, and to use this framework when new processes and systems are designed.

Undoubtedly, as your staff carry out day-to-day activities in their work situations, they become set in their ways of operating and the procedures they follow. Often there is no obvious recognition of the constraints and the opportunities to simplify activity and eliminate waste in your organization's operation. In some respects, this reluctance to change an established way of doing things may be rooted in a fear of job loss.

However, as we have discussed, within the service organizations, leaning out operations results in an organization's ability to free up resources (e.g., both staff time and money). These resources then can be redirected to complete other tasks that are needed, but that have been put on hold due to lack of resources. With the application of lean concepts, methods and tools, there can be substantial gain with respect to

1. how long it takes to complete a process,
2. the number of staff required to complete a process,
3. the reduction in the amount of rework,
4. the reduction in the amount of quality errors,
5. the efficient balancing of labor,
6. reduction in the amount of waiting time, and
7. achievement of outcomes with greater efficiency and effectiveness.

This approach to analyzing processes within the context of intended program outcomes and impact has the power to tap into the latent energy of personnel and to release resources that are trapped in a viscous cycle of wasted work efforts. Furthermore, by applying lean concepts, methods, and tools, your service organization can build its capacity to achieve program goals and improve your accountability for any public dollars you are allocated.

THE POWER OF LEAN TRANSFORMATIONS

To illustrate the power of lean transformations, we have included case studies within three major types of service organizations: social service, government, and educational organizations. An example of a health-care lean case study is given in Chapter 7: Appendix 7B provides a copy of the A3 report at the University of Michigan Health System. Among all the service sectors, health care has spearheaded the application of lean. There are numerous health-care organizations (all along the supply chain) that have instituted lean training and established offices that facilitate lean transformations.

The case studies in this chapter include these:

1. Social Service: South Carolina Center for Childcare Career Development
2. Government: Minnesota State Government
3. Education: Fox Valley Technical College

CASE STUDY 1. SOCIAL SERVICE: SOUTH CAROLINA CENTER FOR CHILDCARE CAREER DEVELOPMENT

Background

Name. South Carolina Center for Childcare Career Development (SC CCCCD)

Location. Greenville, South Carolina

Description of Organization. SC CCCCD is a nonprofit organization with nineteen full-time staff. Since 1993, the Center for Childcare Career Development has administered a credentialing and career development system for early childhood providers in South Carolina. The Center is part of the South Carolina office of First Steps, a comprehensive, results-oriented statewide initiative for improving early childhood development. This initiative, through county partnerships, provides public and private funds for high-quality early childhood development and education services for children and their families to enable children to reach school ready to learn. The Center works closely with and collaborates with the South Carolina Department of Health and Human Services and the Department of Social Services, technical colleges, and other community partners. The Center's mission is to improve the quality of all early childhood programs in South Carolina through the PD of teachers or caregivers.

Implementation of Lean Methodology

From 1996 through 2001, CCCCD developed the statewide training system for childcare providers. In the spring of 2001, the T.E.A.C.H. Early Childhood® South Carolina (SC T.E.A.C.H.) project was implemented. This project is a comprehensive scholarship program that provides the early childhood workforce with access to educational opportunities, including the sixteen technical colleges in South Carolina. The Teacher Education And Compensation Helps (T.E.A.C.H.) Early Childhood® Project was created by the Childcare Services Association in North Carolina in 1990 to address the complex issues of inadequate education, poor compensation, and high turnover within the early care and education workforce. The program is grounded in the principles of a funding partnership among a scholarship recipient, a sponsoring childcare program, and the T.E.A.C.H. Early Childhood® Project. T.E.A.C.H. has a variety of scholarship offerings for caregivers with the overarching goals of increasing caregiver knowledge and skills, increasing compensation, and reducing turnover.

The SC T.E.A.C.H. project rapidly gained popularity among the childcare providers from 2001 through 2007. The growing number of applicants and scholarship recipients made it difficult to operate the program "the way it was always done." The SC T.E.A.C.H. department head describes the existing situation as following (Nodine & Bogatova, 2008):

> Millie [center's executive director] felt as our workload was increasing daily, staff were trying with all their abilities to stay in front of the ball but she felt that we were treading water and just keeping our nose on top of the water . . . we were almost drowning in work. . . . At times I felt terrible watching the frustration of my staff and the work load. . . . Our work load began to triple . . . and our organizational structure remained the same, our processes remained the same . . . we had to do something because we couldn't keep all the balls in the air. I felt a train wreck coming.
>
> Millie saw them [process improvement specialists] at the Professional Development Institute Conference in San Antonio, Texas . . . then again in Maine at the Registry Conference . . . that is where she put two and two together and knew they could help us. She attended sessions on the Post-it® note process. . . . I knew we would bring them to the CCCCD.

To address these challenges, in 2007 CCCCD's executive director made a decision to use lean concepts as a method to increase processing capacity to accommodate the growing number of clients, to improve quality of service, to establish more-efficient and more-effective systems, and to create processes that facilitated the accomplishment of program goals and ensured positive client outcomes or results.

Process Improvement Goals. CCCCD used lean concepts and methods for

1. Decreasing the cycle or lead time to no more than fifteen working days for the following processes:
   a. Reimbursement of the predetermined percentage of cost of tuition, books, and payment of the travel stipend.
   b. Issuance of the credentials indicating a successful achievement of a certain level of education. The Center awards three credentials: the South Carolina Early Childhood Credential, the ABC Family Childcare Credential, and the South Carolina Infant or Toddler Credential.
   c. Issuance of bonuses upon successful completion of the contract year and six months thereafter.
2. Reducing the information deficit or lack of information that prevent work processes from moving forward. For example, when applications come in to the CCCCD without all the required information completed, staff spend time to gather the missing information.
3. Eliminating convoluted pathways, i.e., complicated pathways with many twists and turn that people or material must travel within the flow of a work process. An example of a convoluted pathway is the physical movement of the application through a number of locations (back and forth) for various signatures or additional information before it can be approved.
4. Preventing resource depletion, i.e., when critical resources are allocated (e.g., personnel, time, and money) to complete work activities that add no value to the service delivered or materials produced, thereby diverting these resources from VA work activities. An example is when staff spend a lot of time and effort on processing applications that end up being denied due to ineligibility, resulting

in insufficient time to efficiently respond to eligible clients and provide them a service.

Steps Taken. The vision and consistent and ever-present leadership of the executive director was the key to successful implementation of lean thinking at CCCCD. All staff, from frontline employees to the executive director, attended the introductory process improvement session as a first step to a lean transformation. This session, over two days, provided basic understanding of lean concepts and tools, such as value stream and process flow mapping; VA, NVA, and RNVA activities; URs; root cause analysis; prioritization; and action planning.

The next step in the process of the lean transformation was the collection of data concerning work processes to produce T.E.A.C.H.'s current-state VSM. This data collection required observing the process, defining the steps in the process, and measuring the cycle time of steps in the process. These activities facilitated the understanding of the current state of the T.E.A.C.H. project processes by the team. In addition, each team member was asked to provide thoughts and opinions regarding various aspects of the current processes, such as the time it takes to process an application, reimbursement, credential, and so on; how difficult it is for the staff or clients to follow outlined process; how clear the staff or clients are about the process steps, the reason(s) or rational(s) for doing the process or some of the process steps, and so on. Each person's input was kept confidential by the facilitators, but a summary of the aggregated data was provided to stakeholders involved in the mapping activities.

After sufficient data were collected, the team set aside three days to map and analyze current-state processes, identify URs of the current design and the root causes of the problems or issues, visualize the future state, prioritize improvements, and create an action plan to implement the proposed future-state design (see Figure 9.1). The SC T.E.A.C.H. department head describes the process as follows (Nodine & Bogatova, 2008):

> We began by looking at our major tasks; we wrote each T.E.A.C.H. process on chart paper then determined how much time each process took. Some of the processes we mapped were credentials, the T.E.A.C.H. application process, check requests, travel time to get an active folder from the file room, and how much rework was occurring due to multiple storage units (EXCEL spreadsheets versus T.E.A.C.H. database).
>
> We decided, as a staff, to pick the sub-process that seemed the most cumbersome and then prioritize the processes that needed the most work. Our most "clunky" process was how we awarded Credential/Bonus/and pins. We ultimately all felt we knew where the slowdown was occurring . . . and once we mapped it out, we realized we all held a piece of the process. We were able to laugh out loud as we wrote down each step because it seemed so foolish how much re-work was happening and we somehow were justifying it as internal quality control.

As a result of the new design, several additional lean projects emerged, as described below.

Success Stories

Story 1. One of the most significant improvements that CCCCD achieved was the reduction of the amount of time necessary to process a credential or bonus. Table 9.1 provides data on current state, future state, and percent of improvement.

FIGURE 9.1. Current Value Stream Map of T.E.A.C.H. Process

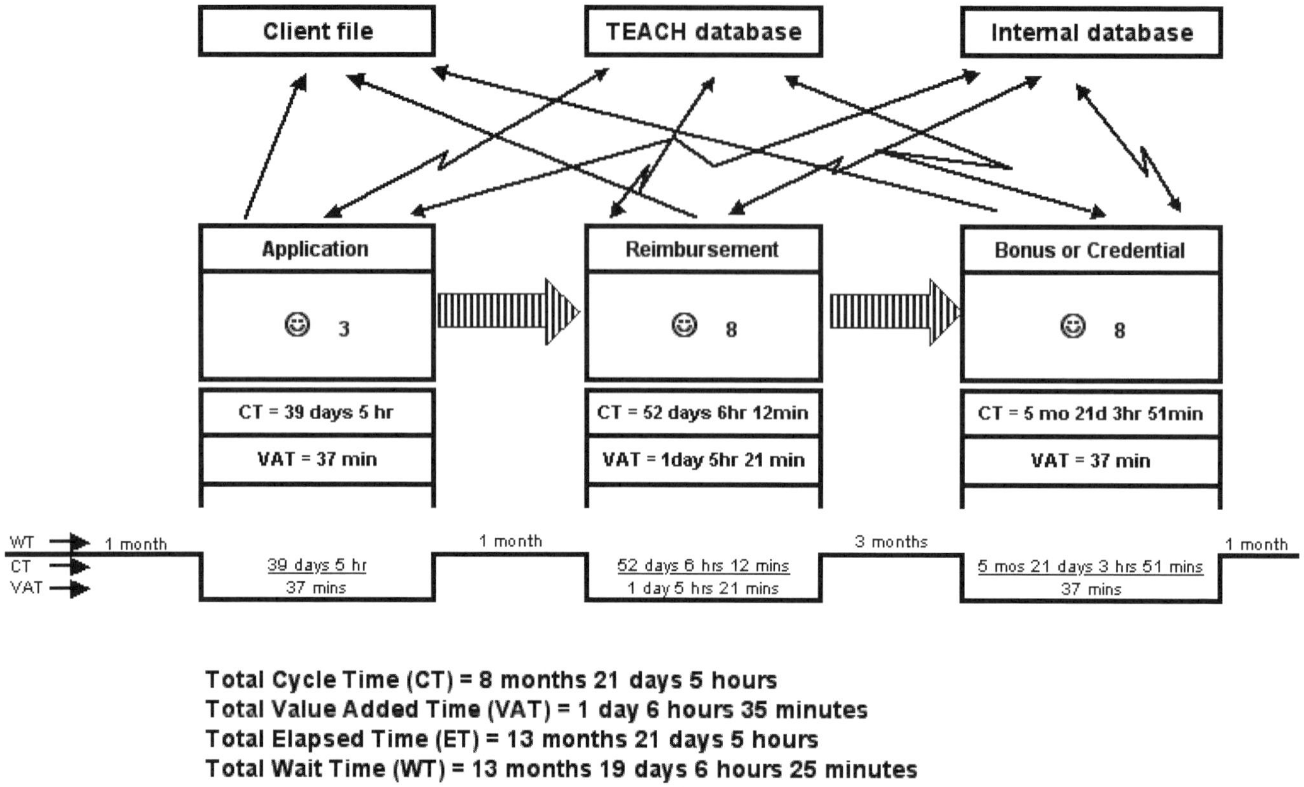

Total Cycle Time (CT) = 8 months 21 days 5 hours
Total Value Added Time (VAT) = 1 day 6 hours 35 minutes
Total Elapsed Time (ET) = 13 months 21 days 5 hours
Total Wait Time (WT) = 13 months 19 days 6 hours 25 minutes

The SC T.E.A.C.H. department head described the results of the implementing lean concepts and tools (Nodine & Bogatova, 2008):

> We redesigned our credential forms. Pre-flow calculations showed that 7 out of 10 forms were filled out incorrectly and post-flow calculations showed 1 out of 10 forms was filled out incorrectly. The participants reported that the new form was more functional.
>
> The time and waste reduction of the credential process is almost embarrassing to talk about. I need you to understand that our checks are processed off site at an accounting firm. Our pre-flow calculations showed it took 73.31 days to process a credential from the day the paperwork entered our office until all paperwork was filed and the case was closed. Ouch, I know. . . . Post flow results yielded 21.72 days. An improvement of 70.37%! This was the one

TABLE 9.1. Credential/Bonus Issuing Process

Metric	Current State	Future State	Improvement
Number of forms with inaccurate information (out of every 10)	7	1	86%
Process time (days)	73	22	70%
Number of process steps	129	17	87%

process that needed the most work. We have virtually removed all the yellow and red dots from this process and we are sticking with a green dot process. It took some getting used to but the ease and speed of the process makes us happy and the "super client" happy.

Recipients are extremely pleased with the speed of the reimbursements. It's so important to me that if a recipient is due a reimbursement that they get their money as soon as possible.

Story 2. Another significant improvement that the Center achieved was a reduction of the storage units that contained service-related information. Table 9.2 provides data on current state, future state, and percent of improvement.

TABLE 9.2. Information Storage Units

Metric	Current State	Future State	Improvement
Number of storage units containing similar information	7	4	43%
Number of storage units containing obsolete information	28	11	61%

The SC T.E.A.C.H. department head described the results as follows (Nodine & Bogatova, 2008):

> Storage Units: I was tracking early childhood credentials in 7 places prior to the mapping process. I think the reason I had established so many storage units (Excel spreadsheets, database, master list, other miscellaneous folders), you know just in case the database didn't pull the report the "right" way. I had to get over it and just trust the T.E.A.C.H. FileMaker Pro database to pull the reports I needed. I figured if I couldn't pull the report then I would call NC [North Carolina] and have someone walk me through what to do.
>
> Shredding Projects: I'm embarrassed to admit that we had never shredded anything since T.E.A.C.H. began its existence in Spring 2001. Imagine the file cabinets filling our work room. We spoke with our funder, CSA, and auditors in regard to how long we needed to hold onto paperwork. We found for auditing purposes that if we had processed an invoice or check request that we need to hold onto paperwork for 3 years, no money processed . . . 6 months.

Story 3. The Center significantly improved the retrieval time of the active client folder. Having a central location for all clients' files resulted in a large amount of travel time accumulated by each staff member. Moving clients' folders into the work area of the staff member responsible for them significantly reduced required travel time. Table 9.3 provides data on current state, future state, and percent of improvement.

TABLE 9.3. Amount of Time Needed to Retrieve an Active Folder

Metric	Current State	Future State	Improvement
Travel time (seconds)	110	15	86%

The SC T.E.A.C.H. department head described the results as follows (Nodine & Bogatova, 2008):

> Filing Cabinets: KeyStone [consultants] hadn't even made it to the airport before we moved all our active T.E.A.C.H. recipient file cabinets into each counselor's office. I spoke with Millie as we walked back to the office and then the move occurred. Ah, the joy! Previous round trip time = 110 seconds and after the move of the cabinets = 15 seconds.

Outcomes

By 2008, CCCCD successfully learned lean methodology and implemented a number of lean projects to meet the growing demand for early childhood PD in South Carolina and dramatically improved cycle time of issuing reimbursements, credentials, and bonuses. The non-value added steps in these processes were identified by the process flow maps and observations. The CCCCD was able to eliminate the waste through process redesign, using the concepts and tools of lean with a focus on one-piece flow.

One of the stakeholders who participated in the lean sessions was a representative of a contracted organization specializing on database design and management. The lean initiative was instrumental in improving communication and understanding between the internal program staff and the external database specialists. The improved communication and understanding translated into a more efficient and accurate database that greatly improved the velocity and accuracy of data-processing functions of the organization.

One of the most significant outcomes has been the change in the staff's way of thinking. The success of lean training is apparent in many areas, but, most important, it changed the way staff think about work processes. Lean thinking and tools are applied in every area of organizational life. Supervisors and staff now ask themselves, "How can this process be improved?" It changed the way staff members think about what they do and how it affects clients. The sense of the empowerment that emerged during the process of redesign was exhilarating. The morale of the organization improved dramatically. The culture began to change and CCCCD transitioned to an organization that continuously learns and improves. Staff members have been empowered to make decisions concerning their work processes and are using sophisticated problem-solving methodology to achieve creative decision making. Every team member is now capable of recognizing that you get what you design.

Lessons Learned

In the words of the SC T.E.A.C.H. department head (Nodine & Bogatova, 2008):

> Don't wait! The longer you wait to change the more difficult it will be to determine changes for the success of your program. If we waited any longer the CCCCD wouldn't be where we are today . . . more effective at reducing waste, happy "super clients," better communication among staff, and empowerment to do the job you have always wanted to do.
>
> Empowerment . . . Earlier I mentioned, "That's the way it had always been done." We made assumptions of things we couldn't do because our funding was federal money, we thought this is what Childcare Association wanted, we thought this is what our funder wanted, we thought this is what our accountant needed, don't auditors need this information? . . .

Staff buy in . . . This was the hardest and the easiest task. The buy-in created an environment that would build on the strengths of each team member.

CASE STUDY 2. GOVERNMENT: MINNESOTA STATE GOVERNMENT

Background

Name. Minnesota's Enterprise Lean Program

Location. Saint Paul, Minnesota

Description of Organization. Enterprise Lean program is a coordinated state government initiative that comes out of the Department of Administration Drive to Excellence program office. The Drive to Excellence program is an ambitious plan for reforming and reshaping the executive branch of state government into an enterprise that is nimble, embraces change, and improves continuously. Under this plan, there are ten currently operating projects or enterprises and five successfully completed projects or enterprises. In an enterprise organization, all components work together seamlessly to realize a common goal, which, in the case of state government, is to serve the citizens of the state. The objectives of the Drive to Excellence program are to increase quality, increase customer service, and reduce the costs of government by

1. creating more "one-stop-shop" opportunities for easier government services,
2. increasing secure and effective electronic delivery of government services,
3. streamlining common functions and activities,
4. optimizing the size of state government by leveraging state worker retirements, and
5. saving money through more efficient, effective service delivery.

The program's focus is on improving organizational performance and results in Minnesota state government agencies. It uses lean concepts and methodologies, Six Sigma tools, and a total quality management approach to establish a culture of continuous improvement and lean thinking in state government to better serve that government's customers and employees.

By year 2015, 50 percent of Minnesota's state workforce will be eligible to retire, resulting in tremendous loss of skills and process knowledge. To prepare for this change in the composition of human capital, the Department of Administration developed a two-fold strategy. One part of the strategy, the "Every Employee Counts" project, addresses workforce development and long-range workforce planning. The other part of the strategy, "Enterprise Lean," focuses on building the capacity of state agencies to redesign and improve their processes. These initiatives are in line with some of the recommendations for Minnesota state government that came from the report, *Grading the States 2008*, produced by the Pew Center on the States in partnership with *Governing Magazine*. This report, prepared in collaboration with a group of academic experts, was the result of Pew's Government Performance Project.

Implementation of Lean Methodology

The proposal to establish Enterprise Lean (Drive to Excellence Program Office, 2008) was presented, accepted, and launched in January 2008. This proposal highlighted the needs and the benefits of implementing lean thinking in Minnesota government and outlined the implementation strategies. At that time, a number of Minnesota state agencies were individually using Six Sigma and lean methodologies. These agencies included Pollution Control Agency, Department of Corrections, Veterans Home Board,

Department of Administration, and Department of Natural Resources. According to the Drive to Excellence program office, main challenges that state government encountered could be described in this way:

> Our workforce is aging. We will be faced with an unprecedented number of retirements in the next few years. There simply will not be enough people in the labor force to fill behind those who will be retiring. We are increasing operations in a competitive environment. Customers' expectations around the quality, timeliness, and in some cases the prices of our services, continue to rise. We are being asked to meet customers' demands with fewer budget dollars. The application of lean principles into our work environment can help address this challenge without undermining the value of the services we deliver to our customers. (Drive to Excellence Program Office, 2008, January, p. 7)

Process Improvement Goals. The main goal of the initiative was to infuse lean thinking principles into the culture of the State of Minnesota Executive Branch. Using lean thinking principles, the state identified eight types of waste to eliminate. These wastes included defects, overproduction, transportation, movement, waiting, overprocessing, inventory, and underutilized resources. Table 9.4 provides specific examples of these wastes the state wanted to eliminate.

TABLE 9.4. Categories and Examples of Waste in the Minnesota State Government

No.	Type of Waste	Examples
1.	Defects	Incorrect data entry
2.	Overproduction	Preparing extra reports, reports not acted upon, multiple copies in data storage
3.	Transportation	Extra steps in process, distance traveled
4.	Movement	Extra steps, extra data entry
5.	Waiting	Processing monthly, not as the work comes in (i.e., financial closings)
6.	Overprocessing	Sign-offs and hand-offs
7.	Inventory	Transactions not processed
8.	Underutilized resources	People doing unchallenging work

Source: From "*Enterprise lean: Business case*," by Drive to Excellence Program Office. Retrieved May 21, 2009, from http://www.lean.state.mn.us/docs/Lean_Business_Case_01.08.pdf

Agency employees were challenged to design and implement more-efficient processes that would require less effort, less inventory, and less time and space, while being highly responsive to their customers.

By implementing lean thinking at the state level, the government expected financial, organizational, operational, and technological benefits. Table 9.5 offers a more detailed look at each type of benefit.

The six-month work plan included the following objectives:

1. To establish an Enterprise Lean Steering Team and develop a business plan.
2. To promote awareness and build understanding among all state agency leaders and managers about the benefits and opportunities that process improvement methodologies provided through high-level lean training.

TABLE 9.5. Benefits of Lean Thinking

Financial Benefits	Organizational Benefits
• Enables cost avoidance • Lowers cost of production and servicing • Enables faster return on investment • Increases cash flow • Increases profitability of products or services • Increases revenue of existing sources	• Improves the ability to serve customers • Builds organization's reputation • Creates new customer opportunities • Fosters vision and mission • Improves employee morale and creativity
Operational Benefits	**Information Technology Benefits**
• Decreases employee workloads for undesirable work • Eliminates NVA activities • Improves internal communication between departments and groups • Improves use of workspace • Increases employee and process productivity • Reduces cycle time • Reduces external inputs to process • Reduces person-hours and process steps • Simplifies processes and workflow steps	• Decreases maintenance or support costs • Improves application or system performance and system utilization rate • Increases efficiency of support activities • Maintains intellectual property investment • Preserves value of technology • Reduces application or system variation (increases reliability) • Reduce paper documentation requirements • Strengthens application or system security • Enables information technology to meet customer expectations for service levels and cost reduction

Source: From "*Enterprise lean: Business case,*" by Drive to Excellence Program Office. Retrieved May 21, 2009, from http://www.lean.state.mn.us/docs/Lean_Business_Case_01.08.pdf

3. To identify and recommend policy and statutory changes that facilitate and support process improvement efforts to eliminate waste, redundancies, duplication of effort, and waiting.
4. To conduct project training within ten agencies.
5. To launch kaizen events within ten agencies.
6. To develop a communication plan.
7. To establish a networking group.
8. To develop student internship program plan for the field of industrial engineering.
9. To create a two- to three-year training plan.

Steps Taken. Multiple steps were taken to achieve the outlined goals. First, a lean office was established and a lean continuous improvement leader was hired to oversee and direct the program.

The second step in the program development was to issue a request for proposal (known as an RFP) and award a contract to a professional lean consultant for six months to assist the Lean Office in development of an implementation and deployment plan, in designing and facilitating training programs and communication processes across the executive branch, and in providing lean master facilitation services for a number of initial projects within the state agencies.

Four pilot lean projects were initiated within the department of administration to test the effectiveness and applicability of lean concepts and methodology. The participants of the pilot projects used value stream mapping to understand their current processes, visualize future improvements, and create an implementation plan. A 5S training program was developed and tested. A number of agencies, including the Department of Health and the Office of Enterprise Technology, identified initial processes for a lean application.

An opportunity for open communication with a number of lean thinking organizations from the private sector was created through the Minnesota Business Partnership. Meetings with lean specialists from Hormel, Ecolab, and General Mills were held to discuss the specifics of implementing lean thinking in large organizations.

In June of 2008, the first issue of E-Lean Update, a monthly electronic newsletter providing updates on how lean is transforming Minnesota state government, was distributed to the interested staff, informing them about lean successes, future events, and training opportunities. At that time, thirteen Minnesota state agencies were involved in a lean transformation. By April 2009, nineteen out of twenty-four cabinet-level state agencies and more than thirty out of eighty other state agencies were involved in the lean transformation, agencies implemented from one to ten kaizen and 5S events (Figure 9.2), 435 employees attended Lean 101 training, 128 employees were trained as kaizen facilitators, and 600 employees participated in kaizen teams (Figure 9.3).

The lean movement is gaining momentum and producing anticipated results. The state government has three main goals it set to achieve by the end of 2010:

1. All cabinet-level state agencies will have implemented systemic and effective process improvement efforts.
2. Executive branch agencies will be able to demonstrate and sustain measurable improvement gains in many key program areas.

FIGURE 9.2. Number of Participants in Lean Events

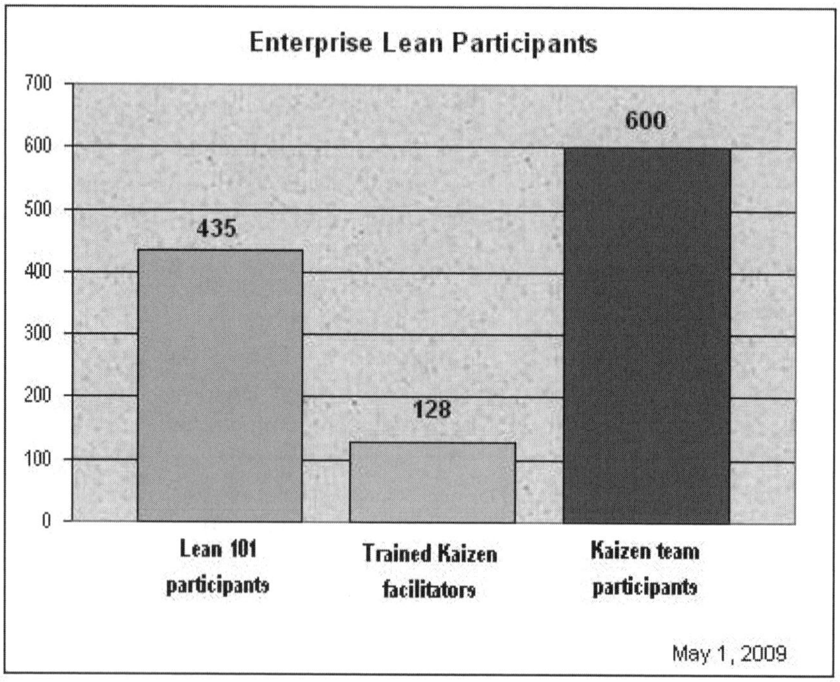

Enterprise Lean Participants

435 — Lean 101 participants
128 — Trained Kaizen facilitators
600 — Kaizen team participants

May 1, 2009

Source: Retrieved on May 21, 2009, from http://www.lean.state.mn.us/index.htm

FIGURE 9.3. Number of Kaizen and 5S Events Held by Agency

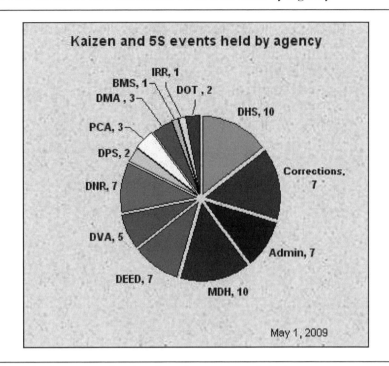

Source: Retrieved on May 21, 2009 from http://www.lean.state.mn.us/index.htm

3. Minnesota will be recognized as a national leader in providing efficient and effective state government services.

Achieving these goals is possible based on many success stories that came to life since the launch of the lean transformation by the Minnesota state government. Some of the success stories are discussed below.

Success Stories

The Department of Human Services' Health Care Operations (HCO) Division is one agency that fully embraced lean and opened itself to a lean transformation. The HCO held five kaizen events since it began the lean implementation in May 2008. The agency improved the lead time on the redesigned processes by an average of 85 percent. However, the impact of the lean transformation extended beyond measurable improvements. Implementing lean helped HCO staff see the existing processes in a different light and challenged employees to think critically about their work processes and how to improve them. Establishing a highly functioning lean steering committee was key to the HCO's success. The committee included division directors, management, and team leaders from the kaizen events. The committee members continued to meet biweekly to sustain the improvements made, navigate roadblocks to improvement, plan for future kaizen events, and encourage daily improvements.

Sarah Patterson, member of the HCO's lean steering committee and the lean manager, explains some the lessons she learned from leading lean events in DHS and offers recommendations to other state agencies trying out lean:

The most important thing I've learned thus far is the power of the untapped creativity of the people doing the work on a day-to-day basis. With the help of various lean tools, the kaizen team members are given the opportunity to come together, discuss the issue, and brainstorm innovative ways to better the process. . . .

My biggest recommendation to other state agencies interested in trying out lean would be to start out by hosting a [kaizen event]. . . . When an agency is able to obtain a "big win" from an initial kaizen event, it will create great interest in lean. After the initial "big win," it is also important to allow lean to grow organically throughout an agency and to ensure lean is the correct tool for each targeted issue rather than try to use lean as an answer to every problem or difficulty within the agency. (Enterprise Lean Program Office, 2008, November, pp. 1–2)

Table 9.6 provides additional examples of improvements made by various state agencies.

TABLE 9.6. Additional Success Stories in Minnesota State Agencies

State Agency	Process	Current State	Future State	Improvement
Department of Health	Issuance of duplicate birth certificate	6 days	<1 day	>83%
Department of Health	Replenishment of lab supplies, stock items	1,104 minutes	32 minutes	97%
Department of Health	Replenishment of lab supplies, nonstock items	2,064 minutes	542.4 minutes	74%
Department of Health	Newborn screening process	52.5 days	23.65 days	55%
Department of Health	Newborn hearing detection process	58.3 days	22.7 days	61%
National Guard	Members' bonus pay for required schooling	>30 days	14 days	>53%
Department of Administration	Federal reimbursement bill processing	3 months	1 month	67%
Department of Human Services	Processing enrollment application for personal care attendants and other home-care providers	54 days	7 days	87%
Department of Human Services	Implementation of new Health Care Eligibility and Access (HCEA) policy	10.6 months	3.7 months	65%
Department of Employment and Economic	Development Process of relocating a Work-Force Center	9–15 months	3-7 months	67%-53%
Department of Public Safety	Part-time contracting service from certification through the execution of the contract	72 days	47 days	35%
Department of Public Safety	Process a grant	20 days	15 days	25%
Department of Corrections	Victim notification of the location and status of a general offender	260 hours	3 hours	99%
Department of Corrections	Victim notification of the location and status of a predatory offender	92 days	16.5 days	82%
Department of Veterans Affairs	Drug administration process	8.32 hours	7 hours	16%

Outcomes

Minnesota state government is successfully moving toward its goal of having all cabinet-level state agencies implement systemic and effective process improvement efforts by year 2010. Currently, nineteen out of twenty-four cabinet level state agencies are involved in the lean transformation efforts. These efforts are transforming government processes by reducing their processing time by 70 percent, on average, and by enhancing quality, efficiency, and effectiveness of government services trough innovation and creativity.

The state, as a whole, has not made a mental shift that translates into a culture of continuous process improvement. There are some great successes in various agencies, where each employee is engaged in lean thinking and is empowered to make necessary process changes to improve the value stream.

Lessons Learned

A number of lessons have emerged from the consistent work aiming to implement lean thinking in Minnesota state government. The most powerful lessons teach the staff that systematic application of lean methodology and tools leads to consistent and significant process improvements. Through this process, all people involved come to understand that transformational leadership is required to take an organization to the next level and implement a culture of continuous process improvement. In the words of Tom Baumann, Continuous Improvement Leader,

> [F]or those agencies and individuals who have invested time in and effort toward learning about and applying lean principles, it is indisputable that major improvements in the efficiency and performance of key processes can be achieved. The results are there for all to see.
>
> And as more work units, divisions and agencies apply Lean principles to their processes, the benefits continue to accrue. But while good things are happening and useful changes are being made and sustained, a bigger goal teases us with larger rewards.
>
> Can we engage and inspire our entire workforce in a systematic effort to continuously improve the thousands of processes within government so that they are as efficient, productive and value-driven as possible? Can we build the organizational commitment and culture that allows each employee in every agency to take their process improvement efforts to the next level?
>
> The short, accurate answer is "yes." But—and this is a huge "but"—taking it to the next level starts with leadership. Leaders define the direction and destination, and together with their staff, establish the best path to get there. In agencies that have realized the consistent benefits of continuous improvements, it is because some leader stuck out their chin, and said, "We can be better than we are right now."
>
> As we look to 2009 and ask ourselves what we need to do to be successful this year, let's look to the leaders. And remember that effective leaders can come from anywhere in an organization. (Enterprise Lean Program Office, 2008, December, pp. 1–2)

CASE STUDY 3. EDUCATION: FOX VALLEY TECHNICAL COLLEGE

Background

Name. Fox Valley Technical College, Payroll Department

Location. Appleton, Wisconsin

Description of Organization. Fox Valley Technical College (FVTC), established in 1967, is a public two-year technical college serving a five-county area (Calumet, Outagamie, Waupaca, Waushara, and Winnebago) in the Fox River Valley region of northeast Wisconsin. It is the third largest of sixteen colleges in the Wisconsin Technical College System. The mission of the college is to educate for employment; that mission is summarized in the value proposition "Knowledge That Works." The college is often chosen by individuals and groups in need of these types of targeted services.

The college offers more than two hundred degree, diploma, and certificate programs in more than seventy-nine occupational areas. It also offers training for twenty apprentice trades. In addition, continuing education, customized training, and technical assistance are offered to a large portion of the employers (about 1,500 to 1,900 annually) in the region, thus equipping more than 20,000 employees with an appropriate knowledge base and skill set. FVTC accounts for 22 percent of all contracted training by Wisconsin technical colleges, or approximately two times the number of trainings provided by any other technical colleges in the state. Academic and technical relevance and rigor of the programs are ensured through close interaction with business, industry, and senior educational institutions.

The college employs 333 full- and part-time instructors, 798 adjunct faculty, 297 full- and part-time support staff, and 108 professional, technical, and administrative staff. Operating budget for fiscal year 2007/2008 was $94.4 million, with 46.6 percent coming from local taxes, 11.5 percent form state funding, 12.3 percent from student tuition and fees, 7 percent from other institutions, and 18.4 percent from federal grants. Contracted training with regional employers contributes about $7.5 million annually.

History of Continuous Improvements. In the mid-1980s, FVTC was one of the first educational institutions in the world to apply quality concepts and tools to education. The College participated in the groundwork of developing the Baldrige Criteria for Education that explored the application of ISO standards to suppliers of education, and assisted the National Quality Forum in defining the body of knowledge for college-level curricula in continuous improvement tools and concepts.

In the early 1990s, FVTC, together with other community colleges, formed the Continuous Quality Improvement Network (CQIN). FVTC was the first national winner of the CQIN Pacesetter Award for Quality and the 2001 Wisconsin Forward Award for quality at the mastery level. Today, the college continues to be an active member of CQIN along with thirty-nine like-minded colleges.

Implementation of Lean Methodology

In 2001, FVTC enrolled in a quality-based accreditation process, known as the Academic Quality Improvement Program (AQIP), administered by the Higher Learning Commission, a commission of the North Central Association of Colleges and Schools" (Fox Valley Technical College, 2006). In 2005, rapid process improvement activities were launched to support enrollment services. These activities were aimed at process redesign and were facilitated by a local business partner with an in-house lean initiative. The project was successful at recording initial wins, however transition of key participants to new roles within the organization had minimized the focus on the initiative. The most recent Quality Checkup Report produced by the Higher Learning Commission (March 2009) reiterated existing opportunities for improvement and stated,

> [The College aims to] identify a few critical processes in each area of the institution. The College noted that it doesn't always have well-defined processes. The College plans to prioritize processes and their articulation and link the processes to results and areas of improvement based on their strategic directions and vision (p. 4).

A request for assistance was submitted by Barbara Kieffer, director of Compensation and Benefits, to FVTC's Lean Performance Center. The Center was established as a fee-based business unit in October 2007 to assist area businesses with lean education and event facilitation to support organizational improvement initiatives. The processing of Adjunct Faculty Teaching Agreements (AFTA) was recommended as a project because of the volume (250 AFTAs processed per pay period), the agreement's complexity (each AFTA varied in terms of hourly rate and number of hours paid), and the amount of manual labor required (four hours per AFTA and an additional twelve hours of processing time for final payroll functions) (Wetzel, 2009). The analysis of current state indicated there were several problems: no clear communication process with adjunct instructors regarding the payroll model for adjunct faculty, no standard work for payroll staff, mileage reimbursement that was a separate process with check requests made at the department level, calculations that were performed manually because automated processes were not trusted, and a system that did not support staffing requirements.

Process Improvement Goals. The goals of lean implementation for processing of AFTAs included these:

1. Reduction of total cycle time to thirty days
2. Elimination of number of change requests
3. Elimination of number of expense checks issued
4. Reduction of mailing expenses
5. Reduction of manual cycle time to 140 minutes
6. Elimination of number of AFTA copies made to track status

Steps Taken. To address existing challenges and discover the opportunities hidden in the current AFTA process, a value stream analysis of payroll process was held in November 2008. Various departments and divisions contributing to the process of issuing AFTAs participated in the value stream activity. During this activity, the eight wastes of lean were reviewed and opportunities for improvement were identified. Among the wastes indentified were underutilized human talent by manually verifying pay rate calculations, the need to wait for signatures while routing contracts, inventory in the form of contracts to be paid, transportation of printed forms and materials, defective entries related to pay rates and change requests, excessive motion related to filing and verifying AFTA information, overproduction related to hiring instructors for classes that did not attain student capacity needed, and the processing waste of printing the contracts when so many inaccuracies existed in the forms.

In addition, the analysis revealed that a number of inconsistencies within the various college departments caused confusion among instructors, rework for many support personnel, and variations in pay models. Fifteen specific issues were identified, but the group chose to focus on three key processes where standardization would have the greatest impact. Those processes were (1) how adjunct instructors were paid for days the college was closed, (2) whether they should be paid for class preparation and, if so, at what load factor, and (3) travel pay.

The value stream activity resulted in identification of two rapid improvement events. The first event addressed the variations and inconsistencies of processing and issuing AFTAs before they reached the payroll department. The second event scope was to map the process redesign for the software upgrade, and the third event was a project to make the enhancements to the software and implement an E-AFTA, or electronic approval process. In addition, the team was challenged with gathering and validating baseline data that were not readily available during the value stream mapping process. The first rapid improvement event was held December 8 through 12, 2008. The process improvement team included members of the payroll team, an information systems expert with the human resources department, several participants from the information services department, three departmental assistants who are responsible for generating and issuing AFTAs and work with hiring managers for the various educational departments, a budget manager responsible for the hiring process, and a representative from a regional center. During this event, three standard processes were created: (1) assigning an instructor to a course, (2) educating staff regarding the new standards, and (3) automating the AFTA process.

The second rapid improvement event provided time for the team to truly analyze and develop a model for enhancing the computerized processing methodologies. This segment included developing an Intranet-based validation system to essentially eliminate the printing and processing of the AFTA form.

The project, completed in May of 2009, completed the software enhancements and validated the process flow through testing.

Success Stories

Information sessions held in the months following the value stream were very well attended and received positive responses. Departmental representatives have been optimistic about anticipated results citing the broken processes and their desire to follow the new standards, which outline a better process for all impacted.

Outcomes

More than $150,000 in savings is anticipated as a result of the changes. Travel pay is being eliminated after it was discovered that only 40 percent of the instructors were being paid mileage. Departments will have discretion for a higher rate of pay to cover mileage in unique situations if specific criteria are met. It is anticipated that $67,500 will be saved if three-fourths of the current expenses are eliminated. In scenarios where travel pay is included as part of the teaching contract, it will be issued with the payroll check. By eliminating the separate check-issuing process, the college anticipates an additional $48,000 in savings.

Preparation Pay is also being eliminated but compensation adjustments are being made through an hourly rate increase for adjunct faculty. Estimated savings of $33,000 represents the difference between current dollars paid out for prep time and the pay increase. Again—less than 50 percent of all adjunct faculty were being given Preparation Pay and there was variation in the amount being granted.

Lessons Learned

While lean improvement strategies have not been fully embraced by the college as a strategic initiative, the methodology for improvement has been embraced at many levels. Executives were participative in debriefs regarding the project and embraced the request for process standardization in areas beyond the payroll offices scope of ownership. As a result, they have developed a viable plan of action. The group involved in the AFTA project has found that process mapping was a very successful way of identifying waste or opportunity within workflows. The map showed how complicated the process had become and how misguided the responsibility was to have payroll team members fix or validate inaccurate pay rates. It also allowed the group to look at the upstream processes and realize the impact of how the information was being presented.

The team truly embraced the concept that the people who do the work are in the best position to redesign it. The team found that the lean improvement methodology allowed people to fix problems with a sense of permanency that they generally did not attain. The format also created an environment that allowed team members to identify and support change related to long-standing issues such as lack of standardization for key benefits for adjunct faculty. Willingness of executive deans to support this process change was a key element in the team's success. Their support in standardizing processes and enforcing consistency will be imperative as the project moves forward.

One of the most valuable elements was the implementation of using the team's creative problem solving prior to making additional capital investments. By evaluating and improving the process before investing in software enhancements, the group created a more viable process.

Kieffer, the process owner and team leader for this initiative, has found that team members now apply lean thinking strategies to many of their daily processes. She indicated that the diversity of the group was incredibly helpful in creating a very powerful change model. Employees who previously went with the flow are now taking a more active role in how they perform daily tasks and are identifying opportunities for improvement. In addition, team participants that may have been perceived as change resistant were actually key contributors and are now strong supporters of the process.

The activity has generated enthusiasm for continued work within the human resources department and within other areas of the college, as well. During the course of the mapping activities, an internal resource was trained to provide facilitation assistance for other requests received within the college. The Lean Performance Center at Fox Valley Technical College continues to be a resource for training employees and providing high-level assistance for lean implementation initiatives.

From Knowledge to Practice

Part IV of this book turns to a discussion of the factors that can either enhance or inhibit your organization's ability to sustain process improvements efforts over time, and offers an in-depth view of some lean transformations in service organizations. The challenges and pitfalls associated with continuous quality improvement are identified as are the factors that are critical in determining the success of lean transformations and the creation of an organizational climate and culture that focuses on organizational learning. In addition, to provide further clarification as to the way in which lean concepts and methods can be applied to service organizations, the final chapter in this part offers three case studies from the social service, education, and government sectors. The following exercises provide you with the opportunity to take the knowledge gained in this section and put it into practice.

> Exercise IV.1. Top Five Challenges
> Exercise IV.2. Intake Process at Childcare
> Resource & Referral Center

In light of our emphasis on learning by doing, the final activity for this book asks you to reflect on the knowledge you have gained and the exercises completed. At that point, you are asked to have a discussion on final thoughts about the application of lean in a service environment and any additional lessons learned that can help you in further attempts to apply this knowledge.

Exercise IV.1. Top Five Challenges

1. Using the ten challenges identified and discussed in Chapter 8, identify what you consider to be the top five challenges that your service organization would face when implementing a lean transformation.

2. Provide a rationale as to why you think these would be the most critical challenges.

3. Identify any additional ways that your organization could face these challenges and prevent them from being a roadblock to a successful lean implementation.

Exercise IV.2. Intake Process at Childcare Resource & Referral Center

Read the scenario below and answer the questions at the end.

Setting the Stage

Roles

- Adult female making a request for a subsidy to help pay for child care (Nadia)

- Adult male intake specialist at Childcare R&R Center (Mike)
- Nadia's infant daughter (Sofie)
- Narrator (Suzanne)
- Mike's coworker (male volunteer from participants)
- Nadia's friend (female volunteer from participants)
- Voice of automated phone service at center (female volunteer from participants)

Props
- Sofie (doll)
- Cell phone for Nadia
- Desk phone for Mike
- Envelopes (9 x 12) for incoming applications (fifty, with paper inside)
- File folders with labels
- Office desk and chairs

Narrator: Nadia and her infant daughter, Sofie, are ready for the day's activity. Sofie is only two months old, and Nadia needs to find a childcare arrangement and be approved for a childcare subsidy so she can return to work in two months. It's 9 a.m. Nadia calls the Childcare Resource & Referral Center.

Nadia: [dials a phone number]

Voice of automated phone service [volunteer]: "You have reached Childcare R&R Center. If you know your party's extension, you may dial it at any time. If you are calling for office hours, please press 1; for questions about eligibility requirements for a childcare subsidy, press 2; for questions about your current application for a subsidy, press 3; for questions about childcare arrangements, press 4; for concerns about your current childcare arrangement, press 5; for all other questions, press 6. If you wish to speak to an operator, press 0 at any time. If you want to return to the main menu, press 7."

Nadia: [presses 1]

Narrator: After seven rings, the voice mail for Mike comes on:

Mike: "I'm unavailable to take your call right now, please leave your name, number, and a brief message. I'll return your call as soon as possible."

Nadia: "Hello, my name is Nadia Bogatova and I want to find out what I have to do to apply for a childcare subsidy. Please call me at 630-839-9521."

Narrator: In the meantime, Mike hears his office phone ring, but he is talking with his coworker so he decides to let the call go to voice mail.

Mike: [talking to coworker (participant volunteer) who is in the same office] "Hey, John, did you watch the football game last night. It was a pretty good game, especially since the Bears beat the Packers by twenty points. I can't wait until this weekend to see the next game."

Narrator: Mike spends another ten minutes talking with his coworker about various things of a personal nature. Finally, he says,

Mike: "Guess I better get to work. I have a pile of applications on my desk. Some of them are more than four weeks old. They just keep piling up and I never seem to have enough hours in the day to process them. I know we're supposed to have them done within ten days. What a joke—I've never been able to do that! So many of them come in without all the supporting documents—then I have to do follow-up to get what I need to determine if they are eligible. It's such a hassle and a waste of my time."

Narrator: Mike spends twenty minutes working at his desk.

Mike: [sits at desk with piles of applications scattered around his desk; he picks up one and begins reading it, but after looking at it, says to his coworker], "Oh, great, here's another one of these applications that isn't complete. I wish these applicants could read instructions and 'get it right' first time around. Now I'm going to have to send a letter to the applicant, asking for the original copies of their pay stubs from the past two months."

Narrator: Mike continues to work processing the applications that are piled on his desk. He never does get to his voice mail that day. The next day, he is tied up in an all day out-of-office meeting, so again he doesn't check his voice mail. Nadia now has waited two days and has not received a call back from the Center. On the third day, she calls again. After going through the voice mail system, she leaves a second message:

Nadia: [in an irritated voice] "My name is Nadia Bogatova. I called and left a message two days ago. I need to apply for a childcare subsidy. I really need to find out how to do this ASAP, since I'm scheduled to go back to work in a few weeks and don't have any arrangements made for child care. Please call me at 630-839-9521."

Narrator: Mike decides to check his voice mail messages today. He has more than twenty-five messages. Seeing this, he says to his coworker,

Mike: "Oh, great. I have a ton of messages to go through. I should have done this sooner. There has to be a better way of doing this work—there's never enough time to get it all done. Messages pile up, applications pile up—the majority of which can't be processed because of missing information—and it seems like I'm forever playing catch-up. On top of that, people get really nasty when they haven't received word about their subsidy. I wish they could see what it's like here in the trenches, then they'd realize that this job isn't so easy. And management is just as bad—they have such unrealistic expectations about timeframes and all the workarounds we have to do to figure out if someone is eligible. Half the time, people don't even meet the basic income levels to qualify. You'd think they would check that first before filling out the paperwork. It just makes our work harder. Not only are we processing lots of applications that don't meet basic eligibility, but then we have to deliver 'rejection' notices to them, which they're never happy about."

Narrator: A week passes, and after playing phone tag for several days, Nadia finally talks with Mike, who informs her that he will mail her a copy of the application. Mike takes another three days before he sends a copy of the application to Nadia. Nadia is now at her home, opening the application she received in the mail. As Nadia opens up the envelope she says,

Nadia: "Finally! I can't believe it has taken this long to get this application. It's already been two weeks since I first called the Center. You would think they would have these applications readily available—like on the Internet or at all the daycare centers."

Narrator: After reading through the application, Nadia says to Sofie,

Nadia: "I can't believe all the information they want—birth certificates, if I have other children, who the father is, pay stubs . . . this is going to take forever to complete and their instructions are so confusing. I'm not sure what they need for documentation. I can't believe they need some formal proof that we're not getting child support from your father. You would think I was applying for a job to be president of the United States."

Narrator: After another five days, Nadia finally completes her application and mails it. When the Center receives the application it has been twenty-six days from the first day in which Nadia called the Center to get information about applying for a subsidy. However, it is a Friday afternoon when the Center receives the application, so it doesn't get delivered to Mike's desk until the following Tuesday (Monday is a holiday). When delivered, it goes to the bottom of the pile of another fifty applications. Based on Mike's average number of applications that get processed every day, it is another two weeks before Mike opens up the application that Nadia sent to the Center.

Mike (opening up Nadia's application): "Ok, here's another one of these applications that doesn't include all the documentation that we need to verify her income and child support payments. Until we have that, we can't determine her eligibility."

Narrator: With that being said, Mike sets the application aside because he has a meeting to go to. When he returns, he goes on to the next packet to process, forgetting to make sure he completes the response letter to Nadia about submitting all required documentation. At the end of the day, Mike straightens up his desk and inadvertently places Nadia's application in another person's file. After another week has passed, Nadia says to a friend of hers,

Nadia: "I still haven't heard from the R&R Center about my childcare subsidy. I started this whole process on October 6 and it is now November 25. I still don't know whether I've been approved. I'm scheduled to return to work right after Thanksgiving. What am I going to do? I don't have any other arrangement for taking care of Sofie. I've checked out a daycare center and they have her on the list to start next Monday, but if I don't get the subsidy I can't afford to pay for it. What do you think I should do?"

Nadia's friend [volunteer]: "My mother can help you out by watching Sofie for a few days, but she can't do it for very long, because she's leaving in mid-December to go to Florida for the winter. So you better call the Center again to find out what's up."

Narrator: The next morning Nadia calls again and hears this message:

Voice of automated phone service: "Our offices are closed for the Thanksgiving holiday. We will be back in our office on Monday, December 1. If you want to leave a message, please press the extension you want and leave your message after the beep."

Narrator: Frustrated, Nadia leaves a message for Mike; while she tries to be polite, it obvious in her tone that she's had it with the inefficiency of this agency. Later, Nadia calls her friend and says,

Nadia: "I can't believe this . . . now I have to wait until Monday to call again. I can't stand it when I have to keep calling a place just to get an answer to a simple question. This agency has a serious problem with customer service. Don't they realize that all I want to do is go back to work so I can support my family? Why do they make this so difficult?"

Narrator: Needless to say, this story doesn't end here. When Nadia receives a phone call from Mike on December 3, she finds out that they have no record of her application and that she'll have to submit another one. Nadia submits another application and finally, on January 5, after another month has passed, she receives notification that she has been approved for a subsidy. It has taken more than three months to have her application processed and approved.

Questions:

1. How realistic to you think this scenario sounds?
   a. Very realistic.
   b. Somewhat realistic.
   c. Somewhat unrealistic.
   d. Very unrealistic.

Provide an explanation as to why you view it as realistic or unrealistic.

2. What is an example of a performance measure that the Center has regarding the processing of applications for subsidy?
   a. Phone messages are to be retrieved and responded to within twenty-four hours of receipt.
   b. Once they submit a complete application, parents should receive notification of approval or disapproval within ten working days.
   c. Ninety-five percent of all the applications for subsidy should be approved.
   d. Not sure.

3. What do you consider an unacceptable result of the way this process played out?
   a. The client (Nadia) was frustrated with the service she received.

b. The processing time took too long.
   c. Both a and b.
   d. Not sure.

4. To what extent to you agree or disagree with this statement?
"The subsidy application process is an unlikely candidate for process improvement because the Center has no control over what the client submits to them."
   a. Strongly agree
   b. Agree
   c. Disagree
   d. Strongly disagree
   e. Not sure

Provide a discussion as to why you agree or disagree with the statement.

5. Using the major types of waste and unacceptable results described in Chapter 3, identify the wasteful activities and unacceptable results that are in this scenario.

6. Using the FAQWOES set of essential factors found in a lean organization, identify and discuss at least two possible solutions to the performance issues depicted in this scenario.

Reflections and Lessons Learned

1. Reflect on the knowledge you have gained and the exercises completed throughout this book. Provide a discussion as to final thoughts about the application of lean in a service environment and any additional lessons learned that can help you in further attempts to apply this knowledge.

Appendix: Logic Models

To have a better understanding of what your organization does and why it does it, you need a level of detail that can be provided by a logic model, which further delineates a theory of change as it relates to a specific program effort within your organization. Logic models provide answers to the "why" question and articulate the connection between your program's inputs and activities (i.e., the defined program intervention or strategy for accomplishing goals) and program results. Logic models have been around since the mid-1960s (Suchman, 1967) and they were promulgated through the United Way's Outcomes Measurement efforts (United Way of America, 1996) and the W. K. Kellogg Foundation's initiative to improve programs (1998).

As a visual representation of the underlying rationale, logic models make explicit the causal factors that are related to a desired condition (i.e., the outcomes that a program wants to achieve for a specific target group). They provide a graphic representation of the linkages between program activities and the changes those activities will produce. They clearly specify a program's set of processes that ultimately contribute to outcomes and the longer-term impact of a program, which are measures of a program's effectiveness. This understanding of a problem's antecedent conditions, root causes, risk factors, or predisposing factors establishes the reasoning behind the design of an organizational program to meet the needs of a target group or to address a social problem (Renger & Titcomb, 2002).

Logic models can be used in a variety of ways. They can be developed prospectively, as an organization begins planning and development of a new program that addresses a particular need or problem. They also can be developed retrospectively for programs that are already operational. As delineated by Rogers (2005), logic models

1. provide a guide to evaluating a program's processes and outcomes by delineating a framework for determining research questions based on the causal path or theory of change that underlies a program effort;
2. provide a common, motivating vision and understanding that has been developed through a collaborative process and make explicit the sequence of inputs, activities, outputs, and outcomes associated with a program; and
3. provide a framework for reporting to funders and senior decision makers on the performance and impact of the program effort, as presented through the output and outcome measures.

At a basic level, there are five key components of a logic model: inputs, activities, outputs, outcomes, and impact. Figure A.1 shows an example of a logic model developed prior to an evaluation of a statewide PD system for early childhood educators. There is greater complexity built into this logic model because the program is statewide and it is not operated by a single organization. Therefore, in this example of a logic

If the whole world followed you, would you be pleased with where you took it?

— Neale Donald Walsch

FIGURE A.1. Logic Model for the Arkansas Early Childhood Professional Development System (AECPDS)

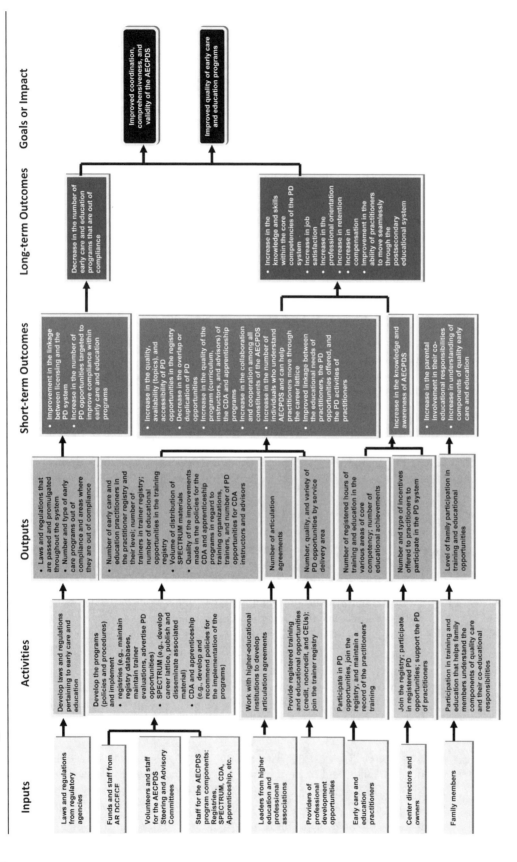

model, we use the term "system" rather than referring to a single organization and a program within the organization.

Inputs. The system's strategy for achieving its long-term goals and associated outcomes begins with the inputs. These inputs are twofold: resources that it needs to function that are dedicated to and consumed by the system (e.g., funds, staff, and participants), and constraints that determine the parameters for the operation of a program.

Inputs for this logic model include the funds from state governing agency, the PD system's steering and advisory committees, the early childhood educators (the targeted participants), the trainers or training organizations, and the early childhood care and education facilities. There also are constraints from the existing legislation and regulations as it relates to childcare standards and number of hours of PD required annually.

Activities. The next component of the logic model specifies the activities, or what is done with the inputs to achieve a mission or long-term goals.

In this model, the activities necessary to establish a statewide PD system and to deliver educational opportunities to early childhood educators, who are the participants, include

1. the development and implementation of a number of organizational components of the PD system, including the policies and procedures of the statewide funding agency and the steering and advisory committees;
2. the participation of early childhood educators in the PD opportunities;
3. the delivery of training by the trainers or training organizations; and
4. the support a facility provides to its early childhood educators.

Outputs. The outputs in the logic model specify the units of service delivered and direct products that result from the activities.

While the outputs specified for this PD system may not be comprehensive, the model does provide a framework for identifying the most salient service units and products delivered. Specifically, the funding agency and governing or advisory bodies developed a number of policies and procedures, completed monitoring and evaluation reports, and addressed a number of complaints within the system. Other advisory committees were responsible for outputs regarding

1. the number of records in practitioner, trainer, and training registries;
2. the number of practitioners served by the PD system;
3. the number of participants (institutions, instructors, advisors, candidates) served by each of the educational programs (e.g., various certificate programs, college degree programs, and noncredit training programs); and
4. the number of articulation agreements between two- and four-year colleges or universities.

For the participating early childhood practitioners, the outputs included the number of various degrees earned by participating early childhood practitioners. Outputs for the trainers or training organizations included the number of PD events or opportunities delivered, the number of verified trainers at various levels, and the distribution of PD opportunities throughout the state's service delivery areas.

The specific outputs for early childhood education facilities are the number of practitioners receiving annual merit increase, the number of paid hours of work for PD, and the number of career counseling sessions provided.

Outcomes. The outcomes of the program can be viewed in terms of short- and long-term outcomes. Outcomes represent the benefits for participants during and after program activities, including changes in the participants' lives, community conditions, or organizational conditions that result from the implementation of an organizational program.

For this early childhood PD system, the short-term outcomes depict the community and organizational changes such as increased availability of high-quality PD opportunities, increased availability of various career pathways, and increased cooperation among all constituents of the system. The long-term outcomes identify the changes in the early childhood educators: an increase in knowledge and skills, an increase in work satisfaction, an increase in retention, and an increase in compensation.

Impact. Ultimately, a program is designed and implemented to achieve a long-term goal, which is the intended impact of a program. Together, the inputs, activities, outputs, and short- and long-term outcomes articulate the theory of change that will result in the ultimate goal of a program.

In the early childhood PD system, the inputs, activities, outputs, and outcomes should lead to its improved coordination, comprehensiveness, validity, and an increase in the quality of care for all young children in the state, which is the long-term goal and intended impact of the early childhood PD system.

Glossary

Accountability: When all members of an organization individually and collectively act to ensure the effective and efficient conduct of activities supporting the achievement of an organization's mission, and provide evidence of such conduct to other individuals within the organization or to outside organizations.

Accreditation: An independent review of an organization's structure and processes to determine the extent to which the organization's performance meets a set of standards established by an independent body.

Action plan: A detailed description of how a process improvement effort will be implemented. It identifies the goal or set of goals and specific objectives of the improvement and how it will be accomplished by specifying the actions and resources needed, making clear who is responsible for each action, the timeline for the actions, and the results of the improvement effort.

Activities: As used in logic models, activities are what is done with inputs (e.g., events, actions) to achieve a mission or program goal(s).

Appreciative inquiry: An approach to evaluation where the focus is on understanding and capitalizing on what is best about a program, organization, or system that will lead to the development of a better future.

Assessment: A judgment made about something based on an understanding of the current situation determined through a data-gathering process.

Auditing: A systematic check or assessment of an organization, especially with regard to its compliance with established standards of practice, typically carried out by an independent assessor.

Balanced scorecard: A one-page tool for translating an organization's strategy into operating terms. It has four columns: Vision, Current Initiative, Business Processes, and Business Results.

Benchmark: A standard against which something can be measured or assessed.

Best practice: The most effective and efficient method of achieving an objective or accomplishing a goal.

Black box: A central component of a system whose functioning is unknown or unclear, for which there are specific inputs and outputs.

Client: A user of the services offered by a social service organization.

Collaborative evaluation: One type of participatory evaluation where the parties working together on the evaluation effort are working together as co-equals.

Command-and-control: A form of management based on the method used in the military, where organizations are top-down hierarchies and managers are expected to make the decisions and manage people through their analysis of budgets, targets, benchmarks, and so on.

Current state of a process: A process map showing the process steps and associated performance measures that is used to identify opportunities for process improvements.

Customer: A current or potential buyer or user of products or services of an individual or organization.

Cycle time: The total amount of time it takes to complete a step within a process.

DMAIC: The Six Sigma problem-solving framework for improving processes, an acronym for Define opportunity, Measure performance, Analyze opportunity, Improve performance, and Control performance.

Effectiveness: The degree to which steps of a process conform to requirements and result in accomplishing established goals and objectives.

Efficiency: The degree to which process steps produce required output and accomplish goals and objectives at minimum resource cost.

Elapsed time: The amount of time it takes to complete a process from beginning to end, to include the cycle time as well as waiting time.

Empowerment evaluation: An approach to evaluation where the concepts, methods, and findings from an evaluation are used to foster improvement and self-determination on the part of the organizations, programs, or people being evaluated.

Evidence-based practice: An intervention to bring about change (improvement) for which systematic empirical research has provided evidence of statistically significant effectiveness as a treatment for a specific problem.

5S: The systematic steps to organize a work environment that include the following: sort, set in order, shine, standardize, and sustain.

Formative (process) evaluation: A type of evaluation used to assess the implementation of a program (particularly its processes) to gain knowledge that can be used to make improvements to those processes.

Future state of a process: A process map showing the process steps and associated performance measures after process improvements have been implemented.

Gemba walks: When organizational managers take regular walks to the front lines where work is completed to look for improvement opportunities through observation and input from the frontline workers.

Impact: As used in logic models, the impact of a program represents the ultimate goal that is to be achieved as a result of a program's effort, consisting of the inputs, activities, outputs, and outcomes.

Inputs: As used in logic models, inputs are the resources dedicated to and consumed by an organization (e.g., funds, staff, and participants).

Jidoka: A Japanese term meaning, "Don't pass along problems to others."

Kaizen event: A focused effort generally lasting three to five days where a dedicated team scopes out a process problem or issue and implements a significant improvement in a process.

Lean thinking: A perspective about process improvement that focuses on delivering the most value to clients or customers while consuming the fewest resources. Value is defined from the vantage point of the client or customer (i.e., identify what the client or customer seeks). Then waste must be eliminated with respect to time, expense, and material at every stage of the operating process to cost-efficiently solve the customer's problems and meet his or her needs so the organization can prosper.

Logic model: A defined framework for understanding an organizational program or service or its theory of change that delineates the connections among inputs, activities, outputs, outcomes, and impact.

Metrics: A standard of measurement of performance, along with the procedures to carry out measurements and to interpret the assessment in the light of previous or comparable assessments.

Mission: An organization's purpose or reason for being. A mission statement accurately explains why an organization exists and what it hopes to achieve in the future, and articulates the organization's essential nature, its values, and its work.

Non-value-added step or activity: Any activity, task, or work within a process that does not contribute to the creation or delivery of a product or service that meets the needs of the client or achieves the expected outcomes, as specified by a program's logic model.

Organizational learning: When organizations continuously modify their actions, values, assumptions, or policies based on knowledge and insights they have acquired through systematic use of data and analysis to identify differences between expected and obtained outcomes of their efforts.

Outcomes: As used in logic models, outcomes are the short- and long-term benefits for participants during and after program activities, defined as a positive change in the participants' attitudes, knowledge, behavior, or status.

Outputs: As used in logic models, outputs are the units of service delivered and direct products that results from program activities.

Participatory evaluation: An approach to evaluation where the program staff and clients (generally referred to as participants) are involved in the planning and implementation of an evaluation.

PDCA: in the process improvement literature, this abbreviation represents the cycle of continuous change, representing the phases of Plan, Do, Check (or Study), and Act.

Performance measure: A number and unit of measure that tells us something important about a product, service, or the processes that produce them. The number gives us a magnitude (how much) and the unit gives the number a meaning (what). Performance measures are always tied to a goal or an objective (the target).

Process variation: When the steps of a process are not clearly defined or followed, resulting in differences in outcome and the way a process is completed, either by different people or by the same person from one time to another.

Program evaluation: A systematic process for gathering and analyzing data to assess an intervention effort implemented to bring about a change in a target group, organization, or society as a whole.

Public agency: An agency of a political division, whether it be at the national, state, or local level.

Push vs. pull: The movement of a product or service through a process. Clients or customers usually pull the goods or service through the process to meet their needs, while suppliers push the goods or service through a process toward clients or customers.

Repetitive processes: Those processes within an organization that are done multiple times, either by different people or repeatedly by the same person.

Required non-value-added step or activity: Any activity or task or work that does not contribute to the creation or delivery of a product or service to the client, but that is required by either internal or external stakeholders.

Rework: When the same activity in a process has to be done more than once due to an error or when the design of a process results in steps being repeated unnecessarily.

Root cause analysis: A method of problem solving aimed at identifying the root causes of problems or events. This analysis is based on the belief that problems are best solved by attempting to correct or eliminate root causes, as opposed to merely addressing the symptoms that are immediately obvious.

Service or product family: The unit of analysis for mapping a value stream. A family consists of a set of services or products that pass through a similar set of process steps from the beginning to the end of a value stream, although there may be some differences in the steps.

Service sector: A division of the economy that is service-producing as opposed to goods-producing. This includes a variety of industries that deliver services as well as material items (sometimes referred to as products), rather than manufacture products.

Six Sigma: A process improvement methodology that seeks to improve the quality of process outputs by identifying and removing the causes of defects (errors) and reducing variation in a process.

Stakeholders: The people or groups who have a stake or vested interest in the operation of a program, including (1) people who have decision-making authority over a program (e.g., funders, policy makers, and advisory groups); (2) people who have direct responsibility over the program's operation (e.g., program developers, administrators, managers, and direct service staff); (3) people who are beneficiaries of a program (e.g., participants, their families, and communities); (4) people disadvantaged by a program, as in lost funding opportunities; and (5) journalists, taxpayers, and the public, particularly for publically funded programs.

Standardization: When work is highly specified as to content, timing, sequence, and outcome.

Standards: A level of quality or excellence that is accepted as the norm or by which actual attainments are judged.

Summative (outcome) evaluation: A type of evaluation where the focus in on the results of a programmatic effort and a judgment is made as to whether a program has achieved its goals.

Takt time: The amount of time to be allocated to the work content so it perfectly matches production time with the demand for products or services.

Targets: A set of performance goals or objectives, expressed in numbers, toward which an organizational effort is directed.

Theory of constraints: An approach to improving a system's performance by identifying and managing correctly any bottleneck, delay, or barrier that inhibits an organization's ability to reach its full potential. Its focus is on improving the flow time of a product or service through a system.

Unacceptable results: The negative consequences of the way work processes are designed and implemented that may be experienced by staff, clients, or stakeholders, causing those individuals to be dissatisfied or frustrated.

Utilization-focused evaluation: An approach to program evaluation that focuses on the utility and actual use of evaluative findings by the intended users, who have the responsibility to apply evaluation findings and implement recommendations.

Value-added time: The amount of time within the cycle time of a process step that is value-added.

Value-added step or value-added activity: Any activity or task or work within a process that contributes to the creation or delivery of a product or service to the client. Dimensions of value include the availability, cost, or performance of a product or service.

Value stream: All the steps in a process from beginning to end required to deliver a service or produce a product for customers.

Wait time: The amount of time between one step of a process and another, during which no activity is being completed.

Waste or wasteful activities: Any activity that consumes resources but creates no value, from the perspective of an organization's client. Forms of waste include waiting, convoluted pathways, rework, information deficits, errors or defects, inefficient workstations, extra processing, stockpiles of materials and supplies, excess services and materials, process variation, and resource depletion.

Worker empowerment: When workers, who are the implementers of work processes, are trained in problem solving and work analysis techniques so they have the skills to continuously find ways to make their work go more smoothly and are proactive in changing their processes accordingly.

Resource Materials

SOFTWARE

eVSM v5. Electronic Value-Stream Mapping. Lean Enterprise Institute, Inc., One Cambridge Center, Cambridge, MA 02142. http://www.lean.org

Metastorm ProVision Business Process Analysis (BPA), Metastorm Enterprise Architecture (EA); Metastorm Business Process Management (BPM) Software. Metastorm Global Headquarters, 500 East Pratt Street, Suite 1250, Baltimore, MD 21202, (443) 874-1300, http://www.metastorm.com/

nFocus Software. The TraxSolutions Outcome Measurement Toolkit (OMT), 6245 N. 24th Parkway Suite 100, Phoenix, AZ 85106, (866) 954-9557, http://nfocus.com/

Process Modeling Software. ProcessModel, Inc., 10602 Covered Bridge Canyon, Spanish Fork, UT 84660, (801) 356-7165, http://www.processmodel.com/index.html

Process Pad Software. Process Master, Ltd. 12 Whitethornes, Mungret, Limerick, Ireland. http://www.processmaster.com/

SurveyMonkey. California Office, 640 Oak Grove Ave., Menlo Park, CA 94025. http://www.surveymonkey.com/

Zoomerang. 150 Spear Street, Suite 600, San Francisco, CA 94105. http://www.zoomerang.com

WEB SITES

American Society for Quality: http://www.asq.org/

Aubrey Daniels International: http://www.aubreydaniels.com/aboutUs/main.asp

Baldrige National Quality Program: http://www.quality.nist.gov/

Balance Scorecard Institute: http://www.balancedscorecard.org/

Benchmarking Best Minds: http://benchmarkingbestminds.com/

Bizmanualz: http://www.bizmanualz.com/

Business Performance Improvement Resources: http://www.bpir.com/

Center for Nonprofit Advancement: http://www.nonprofitadvancement.org/

Drucker Institute: http://www.druckerinstitute.com/

Free Management Library: http://www.managementhelp.org/quality/cont_imp/cont_imp.htm

Harvard Business School (HBS) Executive Education and HBS Working Knowledge e-newsletter: http://www.exed.hbs.edu/

Healthcare Performance Partners, LLC: http://leanhealthcareperformance.com/

Institute for Healthcare Improvement: http://www.ihi.org/ihi

International Organization for Standardization: http://www.iso.org/

Leader to Leader Institute: http://www.leadertoleader.org/about/index.html

Lean Advancement Initiative, Massachusetts Institute of Technology: http://lean.mit.edu/
Lean Education Academic Network: http://www.teachinglean.org
Lean Education Enterprises, Inc.: http://www.leaneducation.com
Lean Enterprise Institute: http://www.lean.org/
Lean Supermarket: http://www.leansupermarket.com/servlet/StoreFront
Management Wisdom.com: http://www.managementwisdom.com/
Minnesota's Enterprise Lean: http://www.lean.state.mn.us/
National Institute of Standards and Technology: http://www.nist.gov/
Next Level Process Partners: http://www.nl-p.com/leantools.html
Polish Six Sigma Society: http://www.polishsixsigmaacademy.pl/en
Six Sigma and Process Excellence: http://www.sixsigmaiq.com
Society for Nonprofit Organizations: http://www.snpo.org
Society for Organizational Learning: http://www.solonline.org/
Team Builders Plus: http://www.teambuildersplus.com/climate_surveys.html
Teambuilding, Inc.: http://www.teambuildinginc.com/
Team Technology: http://www.teamtechnology.co.uk/

ORGANIZATIONS

Academy for Consulting Excellence
N4 W22000 Bluemound Road
Waukesha, WI 53186
Phone: (262) 521-0315
http://www.consultantexcellence.com

American Evaluation Association
6 Sconticut Neck Rd., #290
Fairhaven, MA 02719
Phone: (508) 748-3326
http://www.eval.org

American Society for Quality
P.O. Box 3005
Milwaukee, WI 53201-3005
http://www.asq.org/products/qpress/

The Center for Leadership Excellence
N4 W22000 Bluemound
Waukesha, WI 53186
Phone: (262) 521-0315
http://www.leadexcellence.com

Lean Enterprise Institute, Inc.
One Cambridge Center
Cambridge, MA 02142
Phone: (617) 871-2900
http://www.lean.org/

Peter Barron Stark Companies
11417 West Bernardo Court
San Diego, CA 92127
Phone: (858) 451-3601 or (877) PBS-6468 (toll free)
http://www.pbsconsulting.com/

Quality Resources, Inc.
19321-C US19N
Clearwater, FL 33764
Phone: (727) 669-2242
http://www.qualityresources.com/

Society for Organizational Learning
P.O. Box 381050
Cambridge, MA 02238
Phone: (617) 300-9500
http://www.solonline.org

References

Adler, P. S. (1993). Time-and-motion regained. *Harvard Business Review* (January–February), 97–108.

American Evaluation Association. (2003). *Scientifically based evaluation methods.* Retrieved April 1, 2004, from http://www.eval.org/doestatement.htm

American Society for Quality Buffalo (n.d.). *2010 ASQ Buffalo "Lean Six Sigma" Conference* [Conference Promotional Materials]. Retrieved April 29, 2010, from http://www.asqbuffalo.org/

Argyris, C., & Schön, D. A. (1978). *Organizational learning: A theory of action perspective.* Reading, MA: Addison-Wesley Publishing.

Argyris, C., & Schön, D. A. (1996). *Organizational learning II: Theory, method, and practice.* Reading, MA: Addison-Wesley Publishing.

Armstrong, J. (1982). The value of formal planning for strategic decisions. *Strategic Management Journal, 3,* 197–211.

Barney, H., & Kirby, S. N. (2004). Toyota Production System/lean manufacturing. In B. Stecher & S. N. Kirby (Eds.), *Organizational improvement and accountability: Lessons for education from other sectors* (pp. 35–50). Santa Monica, CA: Rand Corporation.

Bass. G., & Lemmon, P. (1998). *Measuring the measurers: A nonprofit assessment of the Government Performance and Results Act (GPRA).* Nonprofit Sector Research Fund, Aspen Institute, Washington, DC. Retrieved October 19, 2008, from http://www.nonprofitresearch.org/newsletter1531/newsletter_show.htm?doc_id=16677

Bennett, C., & Rockwell, K. (1995). *Targeting outcomes of programs (TOP): An integrated approach to planning and evaluation.* Unpublished manuscript, University of Nebraska, Lincoln.

Bizmanualz. (2008). *What's the difference between process improvement programs?* Retrieved November 11, 2008, from http://www.bizmanualz.com/information/2005/07/06/whats-the-difference-between-process-improvement-programs.html

Butcher, D. (2007). *Is your lean LAME?* Industrial Market Trends. Retrieved June 19, 2008, from http://news.thomasnet.com/IMT/archives/print/2007/07/is_your_lean_la.html

Butler, T., & Synder, G. (2001). Merging the methods of behavior-based safety and continuous improvement: A case study. *The Performance Management Magazine.* Retrieved February 26, 2009, from http://www.pmezine.com/print_article.asp?NID=199

Carman, J. G., Fredericks, K. A., & Introcaso, D. (2008). Government and accountability: Paving the way for nonprofits and evaluation. *New Directions for Evaluation, 119,* 5–12.

Carreira, B. (2004). *Lean manufacturing that works: Powerful tools for dramatically reducing waste and maximizing profits.* New York: AMACOM.

Chapel, J., & Horsch, K. (1998). Interview with Patricia McGinnis. *The Evaluation Exchange, IV*(3/4), 8–9.

Cunningham, J., & Fiume, O. (2003). *Real numbers: Management accounting in a lean organization.* Durham, NC: Managing Times Press.

Deming, W. E. (1986). *Out of crisis.* Cambridge, MA: MIT Press.

Deming, W. E. (1993). *The new economics for industry, government, education.* Cambridge, MA: MIT Press.

Dennis, P. (2006). *Getting the right things done: A leader's guide to planning and execution.* Cambridge, MA: Lean Enterprise Institute.

Dettmer, W. H. (1997). *Goldratt's theory of constraints: A systems approach to continuous improvement.* Milwaukee, WI: ASQ Quality Press.

Drive to Excellence Program Office. (2008, January). *Enterprise lean: Business case.* Retrieved May 21, 2009, from http://www.lean.state.mn.us/docs/Lean_Business_Case_01.08.pdf

Enterprise Lean Program Office. (2008, November). *E-lean update.* [Newsletter]. Retrieved May 21, 2009, from http://www.lean.state.mn.us/docs/E-Lean%20November2.pdf

Enterprise Lean Program Office. (2008, December). *E-lean update.* [Newsletter]. Retrieved May 21, 2009, from http://www.lean.state.mn.us/docs/E-Lean%20December.pdf

Erie County Government. (2008). *Office of Child Support Enforcement Casebuilding Backlog.* Buffalo, NY. Retrieved April 11, 2010, from http://wwww.erie.gov/exec/public/pdf/Office%20of%20Child%20Support%20Enforcement%20Case%20Backlog.pdf

Fetterman, D. M. (1994). Steps of empowerment evaluation: From California to Cape Town. *Evaluation and Program Planning, 17*(3), 305–313.

Fox Valley Technical College. (2006, October). *AQIP Systems Portfolio.* Retrieved May 26, 2009, from http://www.fvtc.edu/public/content.aspx?id=1258&pid=3

Garvin, D. A. (1993). Building a learning organization. *Harvard Business Review* (July–August), 78–91.

Goldratt, E. M. (1994). *It's not luck.* Great Barrington, MA: North River Press.

Green, J. C. (2005). Stakeholders. In S. Mathison (Ed.), *Encyclopedia of evaluation* (pp. 397–398). Thousand Oaks, CA: Sage.

Hagood, C. (2009). *The 12 1/2 truths of a lean transformation.* Retrieved March 14, 2009, from http://leanhealthcareperformance.com/healthcaredocuments/HPPTruthsLeanTransformation.pdf

Hanna, J. (2007). *Bringing "lean" principles to service industries.* Retrieved December 18, 2007, from http://www.exed.hbs.edu/cgi-bin/wk/5741.html

Harrison, M. I. (1994). *Diagnosing organizations: Methods, models, and processes* (2nd ed.). Thousand Oaks, CA: Sage.

Hendricks, M., Plantz, M. C., & Pritchard, K. J. (2008). Measuring outcomes of United Way–funded programs: Expectations and reality. *New Directions for Evaluation, 119,* 13–35.

Higher Learning Commission. (2009, March 26–28). *Quality Checkup Report: Fox Valley Technical College.* Retrieved May 26, 2009, from http://www.fvtc.edu/public/content.aspx?id=1258&pid=3

Hirano, H. (1993). *Putting 5S to work: Step-by-step approach.* Japan: PHP Institute.

Hirano, H. (1995). *5 pillars of the visual workplace.* New York: Productivity Press.

Howardell, D. (n.d.). *Seven skills people need to create a lean enterprise.* Retrieved March 15, 2009, from http://www.rusmart.com/NWIRC-lean/NWIRC-seven-skills.stm

Keyte, B., & Locher, D. (2004). *The complete lean enterprise: Value stream mapping for administrative and office processes.* New York: Productivity Press.

King, J. A. (2005). Participatory evaluation. In S. Mathison (Ed.), *Encyclopedia of evaluation* (pp. 291–294). Thousand Oaks, CA: Sage.

Kirby, S. N. (2004). Malcolm Baldrige National Quality Award Program. In B. Stecher & S. N. Kirby (Eds.), *Organizational improvement and accountability: Lessons for education from other sectors* (pp. 11–33). Santa Monica, CA: Rand Corporation.

Kotter, J. (1995). Leading change: Why transformation efforts fail. *Harvard Business Review* (March–April), 59–67.

Kotter, J. (1998). Winning at change. *Leader to Leader Journal*, 10. Retrieved March 14, 2009, from http://www.leadertoleader.org/knowledgecenter/journal.aspx?Article ID=161

Lean Concepts, LLC. Administrative lean.™ Available at http://leanconcepts.com/case_studies.htm

Lean Education Enterprise, Inc. (n.d.). Proven results where results are most needed: Case study 1. Retrieved April 16, 2010, from http://www.leaneducation.com/case-studies.html#study1

Lean Enterprise Institute. (2008). *A brief history of lean.* Retrieved November 13, 2008, from http://www.lean.org/WhatsLean/History.cfm/

Mann, D. (2005). *Creating a lean culture.* New York: Productivity Press.

Mann, D. (2009). The missing link: Lean leadership. *Frontiers of Health Services Management*, *26*(1), 15–26.

Marchwinski, C., Shook, J., & Schroeder, A. (Eds.). (2008). *Lean lexicon: A graphical glossary for lean thinkers* (4th ed.). Cambridge, MA: Lean Enterprise Institute.

McDaniel, N. (1996). Key stages and common themes in outcome measurement. *Evaluation Exchange*, *2*(3), 1–4. Retrieved October 15, 2008, from http://www.hfrp.org/evaluation/the-evaluation-exchange/issue-archive/results-based-accountability-2/key-stages-and-common-themes-in-outcome-measurement

Metters, R., & Marucheck, A. (2007). Service management: Academic issues and scholarly reflections from operations management researchers. *Decision Sciences*, *38*(2), 195–214.

Miller, A., Simeone, R., & Carnevale, J. (2001). Logic models: A systems tool for performance management. *Evaluation and Program Planning*, *24*, 73–81.

Miller, J. A. [formerly Miller Iutcovich, J. A.) (2002). The politics of unrealistic expectations and the rhetoric of accountability. *Footnotes*, *30*(6), 3.

Mitchell, D. (April 2000). *Review of lean thinking: Banish waste and create wealth in your corporation.* Revised and updated by J. Womack & D. Jones. (2000). Retrieved October 18, 2008, from http://www.amazon.com/review/product/0743249275/ref=dp_top_cm_cr_acr_txt?%5Fencoding=UTF8&showViewpoints=1

Motorola University. (2010). *FAQs: What is Six Sigma?* Retrieved February 20, 2010, from http://www.motorola.com/content.jsp?globalObjectId=3088

National Institute for Standards and Technology (NIST). (2001). *Malcolm Baldrige National Quality Award Program.* Retrieved October 26, 2008, from http://www.quality.nist.gov/

Nave, D. (2002). How to compare Six Sigma, lean, and the theory of constraints. *Quality Progress*, March, 73–78.

Newcomer, K. E. (Ed.). (1997). Using performance measurement to improve public and nonprofit programs. *New Directions for Evaluation*, no. 75. San Francisco: Jossey-Bass.

Nodine D., & Bogatova, T. (2008, April). *Designing and implementing effective and efficient work processes to meet the needs of clients.* Information session presented at the Eighth National T.E.A.C.H. Early Childhood® and Child Care WAGE$® Conference, Chapel Hill, NC.

Office of Management and Budget (OMB). (2004). *Program evaluation: What constitutes strong evidence of program effectiveness?* Retrieved December 9, 2007, from http://www.whitehouse.gov/omb/part/2004_program_eval.pdf

Office of Management and Budget (OMB). (2008). *Government Performance Results Act of 1993.* Retrieved October 18, 2008, from http://www.whitehouse.gov/omb/mgmt-gpra/gplaw2m.html.

Ohno, T. (1988). *The Toyota production system: Beyond large-scale production.* Portland, OR: Productivity Press.

Osada, T. (1991). *The 5 S's: Five keys to a total quality environment.* Tokyo: Asian Productivity Organization.

Owen, J. M. (2005). Learning organization. In S. Mathison (Ed.), *Encyclopedia of evaluation* (pp. 225–226). Thousand Oaks, CA: Sage.

Pannirselvam, G. P., & Ferguson, L. A. (2001). A study of the relationships between the Baldrige categories. *International Journal of Quality and Reliability Management, 18*(1), 14–37.

Patton, M. Q. (1997). *Utilization-focused evaluation: The new century text* (3rd ed.). Thousand Oaks, CA: Sage.

Patton, M. Q. (2002). *Utilization-focused evaluation (U-FE) checklist.* Retrieved October 20, 2008, from http://www.wmich.edu/evalctr/checklists/ufe.pdf

Patton, M. Q. (2005). Utilization-focused evaluation. In S. Mathison (Ed.), *Encyclopedia of evaluation* (pp. 429–432). Thousand Oaks, CA: Sage.

Pew Center on the States. (2008, February). *Grading the states: Minnesota.* Retrieved May 21, 2009, from http://www.pewcenteronthestates.org/uploadedFiles/PEW_WebGuides_MN.pdf

Porche, R. A., Jr. (Ed.). (2006). *Doing more with less: Lean thinking and patient safety in health care.* Oak Brook, IL: Joint Commission Resources.

Powell, T. C. (1995). Total quality management as competitive advantage: A review and empirical study. *Strategic Management Journal, 16*(1), 15–37.

Preskill, H. (2005). Appreciative inquiry. In S. Mathison (Ed.), *Encyclopedia of evaluation* (pp. 18–19). Thousand Oaks, CA: Sage.

Preskill, H., & Russ-Eft, D. (2005). *Building evaluation capacity: 72 activities for teaching and training,* Thousand Oaks, CA: Sage.

Radin, B. (2006). *Challenging the performance movement: Accountability, complexity, and democratic values.* Washington, DC: Georgetown University Press.

Renger, R., & Titcomb, A. (2002). A three-step approach to teaching logic models. *American Journal of Evaluation, 23*(4), 493–503.

Riley, P. (2008). *Ten pitfalls to avoid in process improvement initiatives.* IndustryWeek.com. Retrieved July 25, 2008, from http://www.industryweek.com/PrintArticle.aspx?ArticleID=16593&SectionID=2

Rogers, P. J. (2005). Logic models. In S. Mathison (Ed.), *Encyclopedia of evaluation* (pp. 232–235). Thousand Oaks, CA: Sage.

Rooney, J., & Vanden Heuvel, L. (2004). Root cause analysis for beginners. *Quality Basics*, 45–53.

Ross, J. (1993). *Total quality management: Text, cases, and readings.* Delray Beach, FL: St. Lucie Press.

Rother, M., & Shook, J. (2003). *Learning to see* (version 1.3). Cambridge, MA: Lean Enterprise Institute.

Rush, B., & Ogborne, A. (1991). *Program logic models: Expanding their role and structure for program planning.* Thousand Oaks, CA: Sage.

Senge, P. (1990). *The fifth discipline: The art and practice of the learning organization.* New York: Currency Doubleday.

Shadish, W. R., Jr., Cook, T. D., & Leviton, L. C. (1991). *Foundations of program evaluation.* Thousand Oaks, CA: Sage.

Shewhart, W. A. (1939). *Statistical method from the viewpoint of quality control.* New York: Dover.

Shook, J. (2008). *Managing to learn: Using the A3 management process to solve problems, gain agreement, mentor, and lead.* Cambridge, MA: Lean Enterprise Institute.

Siddons, F. (2010). Buffalo Urban League Six Sigma project: Family preservation services. Buffalo, NY: Buffalo Urban League.

Simon, K. (2009). *The cause-and-effect diagram (a.k.a. Fishbone).* Retrieved March 1, 2009, from http://www.isixsigma.com/library/content/t000827.asp?action=cite& action=cite

Spear, S., & Bowen, H. K. (1999). Decoding the DNA of the Toyota Production System. *Harvard Business Review* (September–October), 97–106.

Spectrum Human Services. (2009). Six Sigma Project: Improve efficiency of attendance verification pros program. Orchard Park, NY: Author.

Spina, M. (2008, March 20). Control board holding back on Six Sigma: Would pay $120,000 instead of $912,000. *Buffalo News*, B1. Retrieved April 29, 2010, from ProQuest Newsstand, Document ID: 1449055191.

Stecher, B., & Kirby, S. N. (Eds.) (2004).*Organizational improvement and accountability: Lessons for education from other sectors.* Santa Monica, CA: Rand Corporation.

Sterman, J. D., Repenning, N. P., & Kofman, R. (1997). Unanticipated side effects of successful quality programs: Exploring a paradox of organizational improvement. *Management Science, 43*(4), 503–521.

Suchman, E. R. (1967). *Evaluative research: Principles and practice in public service and social action programs.* New York: Russell Sage Foundation.

Tapping, D., & Dunn, A. (2006). *Lean office demystified.* Chelsea, MI: MCS Media, Inc.

Trott, P. (2008). *Innovation management and new product development.* Essex, UK: Pearson Education Ltd. Tuckman, Bruce W. (1965) Developmental sequence in small groups, *Psychological Bulletin, 63,* 384–399.

Tuckman, B. W. (1965). Developmental sequence in small groups. *Psychological Bulletin, 63,* 384–399.

United Way of America. (1996). *Measuring program outcomes: A practical approach.* Alexandria, VA: Author.

United Way of Buffalo & Erie County (n.d.). *Six Sigma.* Retrieved on April 30, 2010, from http://www.uwbec.org/content/pages/agencysixsigma

U.S. Department of Labor. (2006). *Government Performance Results Act goals.* Retrieved October 19, 2008, from http://www.doleta.gov/performance/goals/gpra.cfm

U.S. General Accounting Office (GAO; now the Government Accountability Office). (1991). *Management practices, U.S. companies improve performance through quality efforts.* Washington, DC: Author.

Weiss, H. B., & Morrill, W. A. (1998). Useful learning for public action. *The Evaluation Exchange, IV*(3/4), 2–4, 14.

Wetzel C. (2009, May). *Adjunct faculty teaching agreement value steam analysis: Fox Valley Technical College Payroll Department.* Information session presented at the 2009 Lean Educator Conference, Minneapolis, MN.

Winn, B. A., & Cameron, K. S. (1998). Organizational quality: An examination of the Malcolm Baldrige National Quality Framework. *Research in Higher Education, 29*(5), 491–512.

W. K. Kellogg Foundation. (1998). *W. K. Kellogg Foundation evaluation handbook.* Battle Creek, MI: W. K. Kellogg Foundation.

Womack, J. (2006). Value stream mapping. *Manufacturing Engineering, 136*(5), 145–156.

Womack, J., & Jones, D. (1996). Beyond Toyota: How to root out waste and pursue perfection. *Harvard Business Review* (September–October), 140–158.

Womack, J., & Jones, D. (2003). *Lean thinking: Banish waste and create wealth in your corporation* (rev. ed.). New York: Free Press.

Womack, J., Jones, D., & Roos, D. (1990). *The machine that changed the world: The story of lean production.* New York: Harper Perennial.

Index